Exploring the Blue Bioeconomy

This book provides an overview of marine bioresources in the blue bioeconomy. *Exploring the Blue Bioeconomy: Marine Bioresources and Sustainable Applications* delivers potential applications of marine macro- and microorganisms in different industries such as biomedical, functional food, pharmaceutical, cosmeceutical, eco-tourism, industrially important minerals, and enzymes. Aside from the potential industrial applications, the book gives readers an overview of conversion and sustainable utilization techniques for marine bioresources.

Key Features

- Discusses the major sectors associated with the blue bioeconomy and the future potential of each industry for both developed and developing countries.
- Covers the most important aspects of seaweeds with respect to commercialization and conservation, including botanical information.
- Includes a discussion about potential future applications of marine microorganisms in the blue bioeconomy with research highlights.
- Explores the potential industrial applications of marine nematodes and their roles in future agriculture and heavy metal remediation applications.
- Summarizes circular economy principles, upcycling, and recycling approaches to marine litter management under the context of the blue bioeconomy.

The book begins by providing an overview of the blue bioeconomy and then advances through the chapters to discuss potential industries, organisms, and conservation techniques to protect ecosystems and associated organisms from overharvesting and pollution. Thus, readers with any level of educational background can easily understand the content. The text is suitable for undergraduates, postgraduates, industrialists, and policymakers working in the different fields of blue bioeconomy.

Exploring the Blue Bioeconomy
Marine Bioresources and Sustainable Applications

Sanjeewa KKA
Faculty of Technology at the University of
Sri Jayewardenepura, Sri Lanka

CRC Press
Taylor & Francis Group
Boca Raton London New York

CRC Press is an imprint of the
Taylor & Francis Group, an **informa** business

Designed cover image: ShutterStock Images

First edition published 2025
by CRC Press
2385 NW Executive Center Drive, Suite 320, Boca Raton FL 33431

and by CRC Press
4 Park Square, Milton Park, Abingdon, Oxon, OX14 4RN

CRC Press is an imprint of Taylor & Francis Group, LLC

Library of Congress Cataloging-in-Publication Data
Names: Sanjeewa, KKA, author.
Title: Exploring the blue bioeconomy : marine bioresources and sustainable
applications / Sanjeewa KKA.
Description: First edition. | Boca Raton, FL : CRC Press, 2024. | Includes
bibliographical references.
Identifiers: LCCN 2024009460 (print) | LCCN 2024009461 (ebook) |
ISBN 9781032761664 (hardback) | ISBN 9781032761688 (paperback) |
ISBN 9781003477365 (ebook)
Subjects: LCSH: Marine resources. | Marine resources development. |
Sustainable development.
Classification: LCC GC1017 .S36 2024 (print) | LCC GC1017 (ebook) |
DDC 333.91/64--dc23/eng/20240418
LC record available at https://lccn.loc.gov/2024009460
LC ebook record available at https://lccn.loc.gov/2024009461

ISBN: 9781032761664 (hbk)
ISBN: 9781032761688 (pbk)
ISBN: 9781003477365 (ebk)

DOI: 10.1201/9781003477365

Typeset in Times
by KnowledgeWorks Global Ltd.

Contents

Preface ix
Acknowledgments xi
About the Author xiii

Chapter 1   Major Sectors of Blue Bioeconomy 1

1.1   What is Biotechnology? 1
1.2   Blue Bioeconomy on the World Scale 2
1.3   Important Sectors and Industries Associated with
      Marine Bioresources in the Blue Bioeconomy 3
      1.3.1   Fishing Industry 3
      1.3.2   Biomedical Industry 3
      1.3.3   Marine and Coastal Tourism 4
      1.3.4   Marine Renewable Energy 5
      1.3.5   Marine Bioprospecting 8
      1.3.6   Marine Mining and Mineral Resources 10
      1.3.7   Marine Waste Management 12
1.4   Future Directions 12
References 13

Chapter 2   Seaweeds 19

2.1   General Facts about Seaweeds 19
2.2   Botanical Facts about Seaweeds 19
2.3   Reproduction in Seaweeds 21
      2.3.1   Asexual Reproduction 21
      2.3.2   Sexual Reproduction 22
2.4   Classification of Seaweeds 22
2.5   Composition of Seaweeds 22
2.6   Bioactive Compounds Reported from
      Edible Seaweeds 23
      2.6.1   Seaweed Polysaccharides 24
      2.6.2   Seaweed Phlorotannin 30
      2.6.3   Carotenoids 31
      2.6.4   Sterols 32
2.7   Limitations Related to the Bioactive Compounds
      Isolated from Seaweeds 33
      2.7.1   Presence of Heavy Metals 33
      2.7.2   Validation of Bioactive Properties 33
      2.7.3   The Toxic Nature of Solvents and High Cost 33
      2.7.4   Structural Characterization of Purified
              Compounds 33
      2.7.5   Improving the Yield and Characteristics of the
              Seaweed-Based Product 34

| | 2.8 | Future Directions | 34 |
| | | References | 34 |

Chapter 3	Marine Bacteria and Cyanobacteria	39
	3.1	Introduction to Marine Microorganisms	39
	3.2	Major Types of Marine Microorganisms	39
		3.2.1 Marine Bacteria	40
	3.3	Marine Cyanobacteria	51
	3.4	Future Prospectives	52
		References	53

Chapter 4	Marine Fungi	60
	4.1	General Facts about Fungi	60
	4.2	Classifications, Origin, and Habitats of Marine Fungi	60
	4.3	Application of Marine Fungi	61
		4.3.1 Enzymes Isolated from Marine Fungi	61
		4.3.2 Polycyclic Aromatic Hydrocarbons (PAHs) Degradation Properties of Marine Fungi	62
		4.3.3 Heavy Metals Bioabsorption Properties of Marine Fungi	65
		4.3.4 Surface-Active Proteins (Hydrophobins) of Marine Fungi	65
	4.4	Future Directions	67
		References	68

Chapter 5	Marine Microalgae	73
	5.1	Introduction	73
	5.2	Cultivation of Microalgae	74
		5.2.1 Open Systems	74
		5.2.2 Closed Systems	75
	5.3	Microalgae Harvesting Techniques	76
		5.3.1 Harvesting Techniques of Microalgae	77
	5.4	Applications of Marine Microalgae	80
		5.4.1 Applications of Microalgae in the Food Industry	80
		5.4.2 Agricultural Applications	81
		5.4.3 Pharmaceuticals and Cosmetics Applications of Microalgae	82
		5.4.4 Wastewater Treatment Applications of Microalgae	83
		5.4.5 Biofuel Production	84
	5.5	Future Directions	85
		References	85

Chapter 6	Marine Nematodes	92
	6.1	Biodiversity and Taxonomy of Marine Nematodes	92
	6.2	Applications of Marine Nematodes	93

6.2.1   Marine Nematodes in Aquaculture and Fisheries
          Management 93
6.2.2   Marine Nematodes as Ecosystem Architects and
          Emerging Pathogens...94
6.2.3   Effects of Nematodes in the Seafood Industry
          and Aquaculture 95
6.2.4   Marine Nematodes in Environmental Monitoring
          and Bioindication 95
6.2.5   Contribution of Marine Nematodes to Nutrient
          Cycling in Marine Ecosystems 96
6.2.6   Biotechnological Potential of Marine Nematodes 97
6.2.7   Bioremediation Potential of Marine Nematodes
          in Polluted Environments 97
6.3   Future Directions.. 100
References 100

Chapter 7   Applications of Marine Bioresources... 105
7.1   Introduction 105
7.2   Applications of Marine Plant Resources in New
        Plant Fertilizer Products.. 105
        7.2.1   Biostimulation Properties Reported from Marine
                  Plant Fertilizers 106
        7.2.2   Alleviation of Soil Salinity Stress............................ 107
        7.2.3   Aquaponics and Hydroponics Applications............. 107
        7.2.4   Stimulate Root Growth and Nutrient Uptake.......... 108
        7.2.5   Biodegradable Mulching Materials......................... 108
        7.2.6   Disease Resistance and Plant Immunity 108
7.3   Roles of Bioactive Marine Compounds in the
        Functional Food Industry 109
        7.3.1   Antioxidant Properties Reported from
                  Marine Organisms... 110
7.4   Medicinal Value of Marine Bioresources............................. 110
        7.4.1   Antibiotics 111
        7.4.2   Antiviral Properties of Marine Organisms 111
        7.4.3   Anticancer Compounds.. 111
        7.4.4   Anti-Inflammatory Agents....................................... 112
        7.4.5   Cardiovascular Medications..................................... 112
7.5   Cosmeceuticals from Marine Organisms............................. 113
7.6   Applications of Gelatin Separated from Marine
        Fish By-Products 113
        7.6.1   General Extraction Protocol of Fish Gelatin............ 114
        7.6.2   Food Industry Applications of Gelatin..................... 114
        7.6.3   Other Potential Applications of Fish Gelatin 115
7.7   Future Directions... 115
References 116

Chapter 8 Circular Economy Principles Related to the Blue Bioeconomy 122

  8.1   Marine Waste Management and Recycling 122
        8.1.1   Plastic as Marine Contaminant 122
        8.1.2   Marine Plastic Management Strategies 124
  8.2   Upcycling of Marine Resources and By-Products 125
        8.2.1   Principles of Upcycling 125
        8.2.2   Applications of Upcycling across Industries 125
        8.2.3   Environmental Stewardship and Waste
                Reduction with Upcycling Principles 126
        8.2.4   Community Engagement and Education 126
        8.2.5   Challenges Associated with Upcycling and
                Future Outlook 126
  8.3   Sustainability and Marine Biotechnology 126
        8.3.1   Major Sectors can Benefit from Marine
                Biotechnology 127
  8.4   Establishing Circular Supply Chains for Sustainable
        Marine Resource Management 128
        8.4.1   What is Circular Supply Chain? 128
        8.4.2   Key Components of Circular Supply Chains in
                Marine Resource Management 128
        8.4.3   Benefits of Circular Supply Chains in Marine
                Resource Management 130
  8.5   Future Directions 131
  References 131

Index 137

Preface

PURPOSE

The major objective of writing this handbook is to serve as an invaluable companion for students pursuing undergraduate and postgraduate modules in marine bioresources, biotechnology, and related fields and for the general public interested in marine bioresources. *Exploring the Blue Bioeconomy: Marine Bioresources and Sustainable Applications* provides an in-depth understanding of major sectors associated with the blue bioeconomy as well as future potentials of the blue bioeconomy. Furthermore, this book covers recent advances and near-future applications of marine bioresources; thus, it provides updated scientific knowledge and practical applications for its readers.

OVERVIEW

With the increasing global population, events such as global warming and pest and disease outbreaks in traditional land-based agricultural systems, along with disease outbreaks like COVID-19 among human populations, create uncertainty about existing traditional fuel-based economic models and the well-being of our future generations. With these uncertainties, many governments, non-governmental organizations, and policymakers have begun seeking alternative solutions to replace the current unsustainable fuel-based economic model. As a result, bio-based economic models have become a hot topic in numerous global development programs and government agendas.

Among the biobased economic models, the blue bioeconomy is poised to play a significant role in the future. However, there are limited reading materials available on blue bioeconomy-related activities and applications, even though it appears to be a promising alternative to safeguarding the future of mankind. Specifically, undergraduates and postgraduate students studying blue bioeconomy-related applications are required to understand all aspects, applications, and limitations associated with blue bioeconomy-related activities and organisms to gain sound knowledge about the field. It is also important to disseminate essential knowledge and potential associated with this economic model to the public to get more support for future blue bioeconomy-related activities.

Exploring the Blue Bioeconomy: Marine Bioresources and Sustainable Applications covers the most important aspects, applications, and future potentials of the blue bioeconomy. Specifically, the book begins by introducing the blue bioeconomy and the major industries associated with it. This will increase the interest of readers in reading and studying more about the blue bioeconomy. The following chapters then discuss the potential industrial and biotechnological applications of marine macroorganisms and microorganisms with the support of recently cited literature. Specifically, potential applications of seaweeds, microalgae, marine bacteria, marine fungi, and marine nematodes are described in detail. These chapters aim to be beneficial for our future scientists in designing and effectively executing their studies.

The last chapters of this textbook describe the circular economy principles related to the blue bioeconomy. This will enlighten readers to understanding circular economy principles as well as the importance of aspects like recycling and upcycling. In addition, the sustainable use of marine bioresources is also discussed in order to highlight the importance of responsible consumption and utilization of marine bioresources to readers.

As author, I am delighted to recommend this journey into the ocean to study and explore life underwater and the potential applications for the blue bioeconomy. It aims to improve your understanding of the blue bioeconomy and, more importantly, this book will be useful in securing a sustainable blue bioeconomy-based world for future generations.

Sanjeewa KKA

Acknowledgments

I would like to express gratitude to all researchers who work tirelessly to make a better world for future generations.

Sanjeewa KKA

About the Author

Sanjeewa KKA has a BSc honors degree in agriculture from the Rajarata University of Sri Lanka and MSc and doctoral degrees in marine life sciences from Jeju National University, South Korea. In 2019, Dr. Sanjeewa began his postdoctoral research in the Marine Bioresource Technology Laboratory of the Department of Marine Life Sciences at the School of Marine Biomedical Science, Jeju National University in South Korea. Following his postdoctoral research, Dr. Sanjeewa joined the Department of Biosystems Technology at the Faculty of Technology at the University of Sri Jayewardenepura as a senior lecturer in 2021.

Dr. Sanjeewa's research primarily focuses on the bioprospecting of marine organisms in order to discover novel bioactive secondary metabolites for the development of nutraceuticals, cosmeceuticals, and functional foods. His particular interest lies in investigating the anti-inflammatory mechanisms of natural bioactive compounds for treating fine dust-induced inflammatory responses in humans. Dr. Sanjeewa has been recognized among the top 2% of highly cited researchers (awarded by Elsevier BV and Stanford University) in 2021 and 2022, an acknowledgment of his outstanding contributions to the field of marine life sciences.

1 Major Sectors of Blue Bioeconomy

1.1 WHAT IS BIOTECHNOLOGY?

In 1919, the term 'biotechnology' was coined by a Hungarian engineer to describe the use of living organisms to produce products from raw materials (Kennedy 1991). Biotechnology is a diverse segment of science and technology. Biotechnology involves working with living cells and/or molecules derived for applications that benefit human well-being (Kumar 2020). It is a fusion of biological science and engineering, where living organisms, cells, or their parts are utilized to produce products and services. Based on the application or focus, biotechnology can be further classified into different colors. The classification of biotechnology into different colors provides a useful framework for understanding the different areas of biotechnology and their applications (Kafarski 2012). Each color represents a different area of biotechnology, which has led to significant advances in the development of new products, technologies, and therapies. With the continued growth and evolution of biotechnology, it is likely that new colors will emerge, further expanding the scope and potential of this exciting field (Yeung et al. 2019).

The most commonly recognized colors of biotechnology are red, green, and white. Red biotechnology is focused on medical applications and involves the development of new drugs, vaccines, and therapies to diagnose, treat, and prevent diseases (Gartland et al. 2013). Green biotechnology is focused on agricultural applications and involves the development of genetically modified crops, sustainable agriculture, and biofuels (El-Ramady et al. 2022). White biotechnology is focused on industrial applications and involves the development of bio-based materials, chemicals, and fuels (Barcelos et al. 2018; Gartland et al. 2013; Orlandi et al. 2022).

Other than the above-explained colors, several other colors of biotechnology have emerged in recent years such as blue, gray, yellow, and brown. Blue biotechnology is focused on marine applications and involves the study of marine organisms and their potential for use in medicine, biotechnology, and industry. Gray biotechnology is devoted to environmental applications and focused on the maintenance and conservation of biodiversity and removal of pollutants from ecosystems using biotechnological approaches. Furthermore, gray biotechnology approaches are found to reduce the overutilization and pollution of the ecosystem and are using remedial methods to protect the environment (Adepoju, Ivantsova, and Kanwugu 2019). Yellow biotechnology is focused on food and beverage applications and involves the development of new food products, food additives, and bioprocessing (Orlandi et al. 2022). Brown biotechnology is centered around the development of advanced seed varieties that can withstand the harsh conditions of arid regions. This field also encompasses the implementation of innovative agricultural techniques and effective resource management to ensure

TABLE 1.1

Different Colors of Biotechnology

Color	Area of Focus	Example of Application	Reference
Red	Medical biotechnology	Development of vaccines, gene therapies, and diagnostic tests	Gartland et al. (2013)
Green	Agricultural biotechnology	Development of genetically modified crops, biofuels, and environment-friendly products	El-Ramady et al. (2022); Gartland et al. (2013)
White	Industrial biotechnology	Production of chemicals, materials, and energy through biological processes	Gartland et al. (2013); Gavrilescu and Chisti (2005)
Blue	Marine and aquatic biotechnology	Study and exploitation of marine organisms for medicine, food, and energy production	Vieira, Leal, and Calado (2020)
Gray	Environmental applications	Environmental applications involve bioremediation, waste management, and pollution control	Adepoju, Ivantsova, and Kanwugu (2019)
Yellow	Food biotechnology	Improvement of food production, processing, and quality using biotechnology	Orlandi et al. (2022)
Brown	Desert or arid region biotechnology	Development of crops and livestock that can withstand harsh environmental conditions	Galeb et al. (2021)
Purple	Intellectual property rights, ethical issues, and social aspects of biotechnology	Promotion of new biotech inventions, patenting, and study of ethical and social implications of biotech research	Galeb et al. (2021)

sustainable production. Dark biotechnology comprises all scientific research related to microorganisms that have detrimental effects on human welfare and living organisms. This includes activities such as the production of bioweapons, biowarfare, and bioterrorism. Violet/purple biotechnology is unique among the various branches of biotechnology as it is not directly involved in the production of biotechnological products. Its focus is primarily on intellectual property rights and publications in the biotechnology sector. Violet biotechnology encourages the development of new biotechnology inventions and provides patent rights for such inventions. It also delves into the ethical issues and social aspects related to research work in the field of biotechnology (Galeb et al. 2021). Table 1.1 summarizes the major colors of biotechnology.

1.2 BLUE BIOECONOMY ON THE WORLD SCALE

The blue bioeconomy has been growing steadily in recent years, driven by increasing demand for sustainable solutions and the growing awareness of the potential of marine resources. According to the European Marine Board, the blue bioeconomy can be defined as the 'sustainable use of renewable aquatic biological resources, including marine, coastal and inland water resources, to provide food, feed, biobased products, and services.' In addition, the blue bioeconomy sector in Europe had a turnover of approximately €670 million in 2018, and estimates that the blue bio-economy sector in Europe could have a turnover of €1.1 billion by 2025 and €3.7 billion by 2030 (European Marine Board) (European Commission Directorate General for Maritime Affairs and Fisheries 2022). Similarly, a report by the National Oceanic and Atmospheric Administration (NOAA) in the United States estimated that the ocean economy, which includes the blue bioeconomy, had a gross domestic product (GDP) of $373 billion in 2018, with an annual growth rate of around 3.7%. The report projects that the ocean economy will continue to grow in the coming years, with a projected GDP of $446 billion by 2025. However, it is important to note that these estimates may not capture the full extent of the blue bioeconomy, as they only consider certain industries and sectors that are directly related to marine resources. The actual size and growth rate of the blue bioeconomy may be higher if we consider the full range of industries and applications that utilize marine resources.

1.3 IMPORTANT SECTORS AND INDUSTRIES ASSOCIATED WITH MARINE BIORESOURCES IN THE BLUE BIOECONOMY

Marine bioresources are a vital component of blue bioeconomy. They encompass a wide variety of marine organisms, including seaweeds, sea grasses, corals, sponges, invertebrates, vertebrates' bacteria, and fungi. These biological resources are extremely important for a variety of reasons such as food, economy, biomedical applications, biodiversity, and climate regulation. According to the Scopus database (https://www.scopus.com), there are more than 12,500 articles published with the words in title 'marine_organisms' between 1986 and 2022 for different disciplines. Here is the summary for each major segment of marine bioresources.

1.3.1 FISHING INDUSTRY

Humans engaged in fishing before the beginning of written history, utilizing birds' beaks as hooks and plant stalks as lines (Martin 1990). Marine bioresources provide a significant source of food for humans. Fish and shellfish are major sources of protein for many people around the world (Nguyen, Heimann, and Zhang 2020). Fish is a good source of protein, essential fatty acids, minerals, and micronutrients (Sumaila, Bellmann, and Tipping 2016). The fishing industry is a key player in the global economy and also provides employment opportunities for millions of people and generates billions of dollars in revenue (Lam et al. 2020). According to previous reports and studies, wild capture fisheries produce 90 million tons of food each year and according to the United Nations Food and Agriculture Organization, approximately 59.6 million individuals were employed in the primary sector of capture fisheries and aquaculture in 2016 (Anderson et al. 2018; FAO 2018).

1.3.2 BIOMEDICAL INDUSTRY

According to previous studies, secondary metabolites isolated from different marine organisms (metazoans, macroalgae, microalgae, seagrasses, marine bacteria, Archaea, fungi, and viruses) were found to show promising bioactive properties such as antioxidant, antiobesity, anticancer, antihypertensive, antidiabetic, anticoagulant, radioprotective, antimicrobial, anti-inflammatory, anti-Alzheimer's, and hepatoprotective effects (Wijesekara and Kim 2015; Rotter et al. 2021). Even though there are many promising bioactive metabolites available to develop functional products, only a few are commercialized (Camacho, Macedo, and Malcata 2019). The main reason for the slow commercialization of products isolated/identified from marine organisms is the existence of numerous legal and practical challenges. One significant hurdle is the disparity in the pace of progress between technology and legislation. The safety assessments and compliance requirements of products developed from blue biotechnology pose substantial burdens that can delay their entry into the market. Another practical challenge arises from the potential conflicts in spatial utilization. This refers to the conflicts that may arise between existing marine uses, such as tourism or maritime commerce, and the exploration and utilization of marine biota. Resolving these challenges is crucial to facilitating the successful commercialization of marine biotechnology products (Rotter et al. 2021). In addition to the aforementioned properties, the unique genetic and biochemical properties associated with marine organisms make them valuable resources for the development of new products and applications in the fields of functional foods, nutraceuticals, and cosmeceuticals (Asanka, Kim, and Jeon 2018). In the following sections, potential economically important marine organisms will be discussed in detail.

1.3.3 Marine and Coastal Tourism

Coastal and marine tourism represents the most significant sector within the travel industry. It has traditionally been linked to the classic imagery of beach vacations, often characterized as 'Sun, Sand, and Sea' experiences. In recent times, the marine tourism segment has expanded to encompass a wide spectrum of activities related to boating and watercraft, forming what is now referred to as 'Blue Tourism' (Hall 2001; Papageorgiou 2016; Martínez Vázquez, Milán García, and De Pablo Valenciano 2021; Wilks 2021). Additionally, there are also various other leisure water-based activities and nautical sports that take place primarily in coastal waters such as scuba diving, fishing, water skiing, and windsurfing (Diakomihalis 2007). Some of the aforementioned activities are briefly explained in the following sections.

1.3.3.1 Scuba Diving

Exploring underwater environments using scuba gear to observe marine life, coral reefs, and other submerged features is known as scuba driving. Scuba diving has witnessed a surge in popularity, with the Professional Association of Diving Instructors (PADI), representing a significant portion of the global scuba diving market, having issued over 28 million diver certifications worldwide since its inception in 1976 (Kauling et al. 2023; PADI 2021). This remarkable growth attests to the widespread appeal and interest in scuba diving as a recreational and adventure activity (PADI 2021).

1.3.3.2 Spearfishing

The practice of using a spear to catch aquatic organisms is an ancient human activity with a history dating back at least 90,000 years (Yellen et al. 1995).

Spearfishing involves the use of handheld underwater equipment for capturing marine organisms, including fish, cephalopods, and crustaceans. Spearfishing is exclusively performed through free diving, or pole spears, hookah diving, scuba diving, etc. The most commonly used gear for underwater harvesting includes spearguns and Hawaiian slings (Sbragaglia et al. 2023). Spearfishing is also a highly selective fishing method; participants see and target individual fish that may be endangered. This might negatively impact the health and sustainability of a particular ecosystem. Thus, it is essential to monitor spearfishing if local authorities are to promote spearfishing as a part of marine tourism (Young, Foale, and Bellwood 2015).

1.3.3.3 Water Skiing

Water skiing is generally considered as a luxury sport and is a dynamic and captivating water sport with a rich history, diverse techniques, and increasing recognition as a competitive discipline (Bray-Miners, Runciman, and Monteith 2012). Water skiing's history dates back to the early 20th century when Ralph Samuelson became the first person to successfully ski on the water in 1922 on Lake Pepin, Minnesota (Runciman 2011). With time water skiing gained popularity as both a recreational activity and a competitive sport. The sport continued to evolve, with advancements in equipment, techniques, and the establishment of international governing bodies like the International Waterski and Wakeboard Federation (more at https://iwwf.sport/).

1.3.3.4 Tours to Maritime Parks

People participate in guided tours to protected marine areas, such as marine parks, reserves, or sanctuaries, to observe and learn about the local ecosystems and wildlife.

1.3.3.5 Wildlife Mammal Watching

Participants engage in activities focused on observing marine mammals in their natural habitats, such as dolphins, whales, and seals.

1.3.3.6 Coastal Tourism

Coastal tourism is a specific form of tourism where the predominant attraction and advantage lie in the water or sea element. It is closely related to marine or maritime tourism as it includes activities taking place in coastal waters. However, coastal tourism also encompasses beach-based tourism and recreational activities that occur along the coastline. Examples of such activities are swimming, sunbathing, coastal walks, and other leisure pursuits (Diakomihalis 2007; Karani and Failler 2020).

1.3.4 Marine Renewable Energy

Marine renewable energies offer a promising pathway toward a cleaner energy future (Shields et al. 2011; Zhang, Lin, and Liu 2014). The marine environment offers abundant renewable energy potential, and there is a growing interest in extracting energy from areas with characteristics such as large tidal ranges, fast water flow with or without wave interaction, and significant wave resources (Graziano et al. 2017; Shields et al. 2011). This increasing focus on marine renewables aims to reduce reliance on fossil fuels, diversify energy sources, and alleviate greenhouse gas emissions (Bucher and Bryden 2016; Shields et al. 2011). According to a study by Wang et al. (2019), it has been estimated that the marine renewable energy resources available worldwide surpass both our current and future projected energy demands. However, despite this potential, marine renewable energy currently contributes only a minimal proportion to our overall energy system, accounting for less than 3% of the total (Wang et al. 2019). Taken together, marine renewable energy holds significant potential for development and is expected to play a crucial role in establishing a sustainable energy infrastructure for the future (Miller et al. 2013; Yang et al. 2019). Marine renewable energy encompasses various sources, including offshore wind, offshore solar, and ocean renewable energy, which collectively offer a diverse range of options for harnessing renewable energy from the marine environment. By further exploring and harnessing these marine renewable energy sources, we can make significant strides toward achieving a more sustainable and environment-friendly energy system (Miller et al. 2013; Wang et al. 2019; Yang et al. 2019). The following section summarizes the dominating marine renewable energy generation technologies.

1.3.4.1 Offshore Wind

Offshore wind is utilized in generating electricity and auxiliary propulsion for ships. In the context of generating electricity, offshore wind involves capturing the kinetic energy of the air through wind turbines and converting it into electrical energy (Wang et al. 2019). This process harnesses the power of offshore winds to produce electricity, which can be fed into the grid and used to power homes, businesses, and industries (Esteban and Leary 2012). On the other hand, auxiliary propulsion refers to utilizing the force of offshore winds to directly propel ships, reducing their reliance on conventional fuel and thus lowering fuel consumption. This approach helps improve the energy efficiency of maritime transportation and contributes to reducing carbon emissions in the shipping industry (Li, Wang, and Sun 2020).

1.3.4.2 Offshore Solar Energy

Offshore solar energy can be used for heating and photovoltaic generation on ships, offshore platforms, and islands. Solar thermal systems can provide hot water and space heating, while photovoltaic systems can generate electricity from sunlight. These applications contribute to energy efficiency and sustainable power generation in offshore environments (Kumar, Shrivastava, and Untawale 2015).

1.3.4.3 Tidal Range Energy

Tidal power plants have advantages similar to those of hydroelectric power, including a long lifespan compared to other renewable energy technologies and relatively low operation and maintenance costs (Curto, Franzitta, and Guercio 2021; Harcourt, Angeloudis, and Piggott 2019). Tidal range power is generated by utilizing the difference in water levels between two bodies of water. To create this difference, a barrier or wall is constructed to separate the two areas. As the tide flows in or out, the wall restricts the movement of water, causing a buildup of water on one side and a lower water level on the other side. This difference in water levels, known as the head difference, is crucial for generating energy. When the head difference reaches an optimal level, the water is allowed to pass through the barrage or dam. Within the wall, turbines are strategically placed to capture the kinetic energy of the flowing water. As the water passes through the turbines, it spins the blades, generating mechanical energy that is then converted into electrical energy by a generator (Waters and Aggidis 2016).

1.3.4.4 Marine Current Energy

Marine tidal currents hold immense potential for electric power generation. Marine tidal currents are recognized as a valuable resource for sustainable electricity production (Bahaj and Myers 2003). Marine currents are generated by the movement of tides and ocean circulation, and they can also be influenced by factors such as river outflows and variations in temperature and salinity levels. The kinetic energy present in these marine currents can be captured using different technologies. The principals involved are similar to those of wind energy, where the power available at a specific location is determined by the density of the fluid and the cube of its velocity (Nachtane et al. 2020; Rourke, Boyle, and Reynolds 2010). The main distinction between the two resources lies in the density of the working fluid. Seawater has a much higher density than air (approximately 832 times higher), resulting in higher power output from marine current energy devices compared to wind energy devices of similar size, assuming similar fluid velocities (Rourke, Boyle, and Reynolds 2010).

1.3.4.5 Ocean Wave/Current Energy

Ocean currents are complex phenomena influenced by various factors such as temperature, salinity, wind, tides, and Earth's rotation. They exhibit both periodic and aperiodic movements, including seasonal and short-duration changes, as well as oscillatory and sporadic movements (Crouch et al. 2008; Hsu et al. 2021). Observing and understanding these currents pose significant challenges. The primary drivers of marine currents are temperature and salinity differences, in addition to the influence of wind, tides, and Earth's rotation. Wind and tides are particularly responsible for generating the most intense marine currents (Bahaj and Myers 2003; Crouch et al. 2008; Hsu et al. 2021). When a current is caused by tides, it is often referred to as a tidal stream. Tidal streams are one of the most utilized forms of marine current energy (Esteban et al. 2019). Bathymetry, which refers to the study of underwater depth variations, can also play a role in enhancing tidal streams (Christ and Wernli 2014). In areas where the bathymetry narrows, the current velocity can increase significantly, leading to more favorable conditions for harnessing tidal stream energy. Tidal stream energy has a high potential as a renewable energy source, estimated at around 800 TWh/year. However, to harness this energy, the speed of the currents needs to exceed 2 m/s (Charlier and Finkl 2009).

1.3.4.6 Biomass Energy

Biomass refers to any organic material derived from plants, including algae, trees, and crops. It encompasses both terrestrial and aquatic vegetation, as well as organic waste materials. Biomass is produced through photosynthesis, where green plants convert sunlight into plant material. It can be considered as organic matter that stores the energy from sunlight within its chemical bonds. When the bonds between carbon, hydrogen, and oxygen molecules in biomass are broken through processes such as digestion, combustion, or decomposition, the stored chemical energy is released. This energy can be harnessed for various purposes (Esteban et al. 2019; McKendry 2002). Marine plant biomass has advantages over land-based plants as a biofuel source. Recent advances in using seaweed biomass, through metabolic engineering, to produce liquid biofuels like bioethanol show promise. Seaweed biomass has potential as a relatively unexplored but promising option for biofuel production (Wei, Quarterman, and Jin 2013; Xin, Zhong, and Wang 2023).

1.3.4.7   Ocean Thermal Energy

Ocean thermal energy conversion is a renewable energy technology with a global theoretical potential of up to 30 TW (Langer, Quist, and Blok 2020). The vertical distribution of ocean temperature exhibits two distinct areas with a significant temperature contrast. This temperature difference between the marine surface and depths of 1,000 m can be harnessed for energy production, known as ocean thermal energy (Esteban et al. 2019; Liu et al. 2020). This form of energy harvesting relies on the disparity between warm surface water and deep cold water, which facilitates the operation of a heat engine to generate electricity. The process involves utilizing the heat from the shallow ocean water to warm a liquid within the engine that possesses a low boiling point. As the liquid evaporates, it drives a turbine, generating electricity. The vapor then cools down when it comes into contact with the deep cold water, initiating a new cycle of power generation. This continuous cycle allows for the sustainable production of electricity using the temperature gradient in the ocean (Langer, Quist, and Blok 2020; Liu et al. 2020; Vega 2002).

1.3.4.8   Salinity Gradient Energy

Salinity gradient energy refers to the energy released between two bodies of water with different salt concentrations. Osmotic energy harnesses the difference in osmotic pressure resulting from the contrast in salinity between freshwater and saltwater. When these two fluids come into contact, a balance in salt concentration occurs. To capture osmotic energy effectively, it is crucial to position facilities at river mouths where the salinity difference is significant. However, the installation of osmotic energy facilities at river mouths may face limitations due to potential conflicts with other uses or activities commonly carried out in those areas (Helfer, Lemckert, and Anissimov 2014).

1.3.5   Marine Bioprospecting

There is a growing demand for new antibiotics, chemotherapeutic agents, and agrochemicals that are highly effective, safe with minimal toxicity, and environment-friendly (Hosseini et al. 2022). This need arises from the development of resistance in disease-causing microorganisms, the emergence of naturally resistant organisms, and the rise of highly pathogenic diseases like COVID-19 further emphasizing the urgency to discover and develop new drugs to combat them (Jain and Tailor 2020). In addition to targeting specific diseases like AIDS, there is a need for new therapies to treat secondary infections that arise due to weakened immune systems, particularly in immunocompromised individuals such as cancer and organ transplant patients (Strobel and Daisy 2003). Additionally, the removal of synthetic agricultural agents from the market due to safety and environmental concerns necessitates the exploration of alternative methods to control pests and pathogens in agriculture. Exploring novel natural products and the organisms that produce them holds great potential for innovation in drug and agrochemical discovery. These are some examples that highlight the requirement of bioprospecting (Strobel and Daisy 2003).

According to the different definitions given by different bodies, collectively bioprospecting can be defined as a 'systematic and organized search for useful products derived from bioresources including plants, microorganisms, animals, etc., that can be developed further for commercialization and overall benefits of the society' (Oyemitan 2017). Marine ecosystems are a promising source for bioprospecting as these ecosystems cover a large portion of Earth compared to terrestrial ecosystems (Manikkam et al. 2019). In recent years, the exploration of marine natural products has led to the discovery of numerous drug candidates (Leal et al. 2012; Sanjeewa et al. 2021). While many of these molecules are still in the preclinical or early clinical development stages, some have already reached the market, such as cytarabine, Prialt®, and Yondelis® (Rocha et al. 2011). Studying the ecology of marine natural products has revealed that many of these compounds act as chemical defenses and have evolved to be highly potent inhibitors of physiological processes in the prey, predators, or competitors of the marine organisms that produce them (Haefner 2003). Other than the bioactive compounds isolated from marine microorganisms, it has been discovered that some natural products isolated from marine microorganisms are either proven or suspected to have a microbial origin. In fact, it is now believed that the majority of these molecules are derived from marine microorganisms, which possess immense genetic and biochemical diversity that is only just beginning to be understood (Goodfellow and Fiedler 2010; Haefner 2003; Leal et al. 2012). This suggests that marine organisms have the potential to be a valuable source of novel chemical compounds for the development of more effective drugs to treat different disease conditions.

Basically, bioprospecting involves a methodical exploration of wild resources to discover genes, natural compounds, and organisms that can be used to develop products using bio-physicochemical and genetic approaches while preserving nature and its resources (Figure 1.1) (Tiwari and Chauhan 2021). Thus, bioprospecting includes three major areas – chemical prospecting, gene prospecting, and bionic prospecting – described in the following sections.

1.3.5.1 Chemical Prospecting

Chemical prospecting involves the search for natural compounds with potential applications in various fields, such as pharmaceuticals, cosmetics, and agriculture. The goal is to identify chemical substances from natural sources that possess beneficial properties (Eisner 1994; Roberts 1992). More than 60% of pharmaceuticals are related to natural products, and chemicals produced by

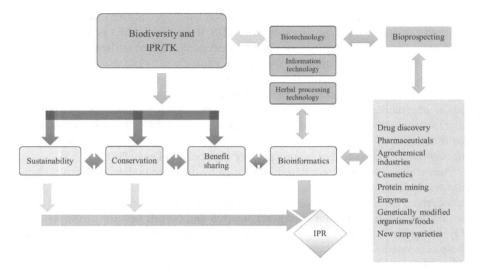

FIGURE 1.1 Major areas of bioprospecting and applications. Abbreviations: IPR: intellectual property rights; TK: traditional knowledge.

living organisms. Despite this, the rate of natural product discovery has slowed over the past few decades (Kim et al. 2021). As mentioned earlier, the chemical diversity of marine organisms is yet to be explored. Specifically, well-planned marine chemical prospecting projects have the potential to reveal promising drugs to treat various diseases such as cancer and chronic inflammatory diseases (Leal et al. 2012).

1.3.5.2   Gene Prospecting

In gene prospecting, the focus is on searching for genes or genetic material that can be utilized for various purposes, including gene therapy, biotechnology, and the development of novel traits in crops or organisms. The aim is to identify genetic resources that can be harnessed for specific applications.

1.3.5.3   Bionic Prospecting

Bionic prospecting involves studying and drawing inspiration from natural organisms to develop innovative products or technologies. By observing the structures, functions, and behaviors of organisms, scientists can create bioinspired designs that mimic nature's efficiency and effectiveness.

1.3.6   Marine Mining and Mineral Resources

The mining industry is witnessing a global trend where terrestrial resources are becoming more difficult to exploit, leading to rising costs for minerals and metals.

As a result, there is a growing interest in exploring the ocean as a potential source of valuable resources. While the deposits in the ocean are generally smaller in quantity, they often contain high-value metals. For instance, seafloor massive sulfide deposits found in the ocean can hold valuable minerals like copper, lead, zinc, and sometimes even gold and silver. Despite the challenges associated with marine mining, the allure of these high-value metals makes it an attractive option for the industry (Teague, Allen, and Scott 2018). The extraction of metal resources from the ocean is a crucial aspect of ocean engineering. Marine mining involves exploring, extracting, and processing valuable minerals and resources found in the seabed (Rona 2008). This industry has gained prominence due to the growing global demand for metals and minerals, along with advancements in mining technologies (Ludvigsen and Sørensen 2016). However, marine mining also brings forth substantial environmental and social challenges.

Marine carbonate sediments, such as those found in the Brazilian Exclusive Economic Zone in the tropical Southwestern Atlantic Ocean, have significant economic value due to their mineral composition. These sediments are rich in calcium minerals and valuable trace elements, making them highly appealing to various industries. In fact, the Brazilian Exclusive Economic Zone holds the largest known deposit of marine limestone worldwide. The estimated reserves in this area exceed 135.5 million tons of calcium carbonate ($CaCO_3$). This resource is particularly valuable for applications in agriculture and animal nutrition, making it a highly sought-after supply for these industries (Paiva et al. 2023). Marine carbonate sediments result from the accumulation of sand and gravel derived from various sources such as calcareous algae, algal nodules, corals, mollusks, foraminifera, and benthic bryozoans (Paiva et al. 2023). These sources contain substantial amounts of calcium carbonates, magnesium, and other significant trace elements. Marine mining is an emerging industry that aims to extract valuable minerals and resources from the ocean floor.

While it offers opportunities for accessing untapped resources, marine mining also presents significant environmental challenges. The primary concerns surrounding deep-sea mining revolve around its potential environmental impacts. Currently, the ecological effects of mining rare earth elements in deep-sea ecosystems are not well understood and require further evaluation. Additionally, conflicts can arise between marine mining activities and various stakeholders, including fisheries, owners of communication cables, offshore wind farms, and tourism (Ludvigsen and Sørensen 2016).

The global ocean plays a vital role as a source of food, energy, raw materials, clean water, and ecosystem services (Sala et al. 2021). Unfortunately, it faces significant challenges from multiple anthropogenic stressors. To develop a sustainable blue economy strategy, a comprehensive understanding of the environmental impacts is crucial. This includes protecting vulnerable areas, adopting new technologies for deep-sea mineral exploration and mining, implementing marine spatial planning, and establishing a regulatory framework for minerals extraction. Through these measures, we can work toward the sustainable management and utilization of our oceans while minimizing adverse impacts.

1.3.7 MARINE WASTE MANAGEMENT

The problem of the presence of waste in the marine environment has recently taken on the dimensions of a complex and global challenge (Paolo et al. 2020; Tsai et al. 2021). In general, marine debris refers to long-lasting pollutants that enter water bodies, including both freshwater and marine environments (Fauziah et al. 2021). According to previous studies, marine wastes can have various negative impacts on marine ecosystems, including degradation or destruction of particular ecosystems. This is primarily caused by physical intrusion, such as obstructing sunlight, surface scoring, and abrasion. For example, marine debris like plastic bags, fabric, or sheeting can abrade and smother corals. The consequences of marine litter extend beyond environmental concerns and also affect areas such as tourism, economy, safety, health, and culture. One significant issue associated with most marine waste is the slow decomposition rate, resulting in a gradual but substantial accumulation in coastal and marine environments (McNicholas and Cotton 2019; Prabhakaran et al. 2013).

Plastic debris has received significant attention among all types of marine debris, particularly since the 1970s when plastic production started to escalate. Due to their low decomposition rate in the marine environment, plastics accumulate gradually but extensively (Chen et al. 2019). Unfortunately, a substantial amount of plastic continues to be discharged into the marine environment (Derraik 2002; McNicholas and Cotton 2019). In 2010, it was estimated that 4.8–12.7 million metric tons of 'mismanaged' plastic waste enter the ocean annually (192 coastal countries). This includes not only land-based plastics but also abandoned or lost fishing gear, such as fishing nets and lines (Jambeck et al. 2015).

Marine waste management plays a crucial role in the blue bioeconomy, encompassing the sustainable utilization of marine resources. It is essential to implement well-planned waste management programs in order to safeguard sensitive marine ecosystems and support industries connected to the marine environment, such as tourism. Effective waste management strategies can help prevent pollution, preserve biodiversity, and maintain the overall health of marine ecosystems. By addressing marine waste, we can ensure the long-term viability of industries reliant on the marine environment and foster a sustainable balance between economic development and environmental conservation.

1.4 FUTURE DIRECTIONS

Marine biodiversity is a key segment of the blue bioeconomy that depends on and influences the climate of prevailing ecosystems, the quality of the water, and many other ocean state variables. It is also at the core of ecosystem services that can make or break economic development in any region (Estes et al. 2021). Moreover, as we progress into 2030 and beyond, the expanding global population will increasingly rely on marine organisms. The marine biodiversity and related ecosystems can be appropriately harnessed to promote ecosystem vitality, economic productivity, operational efficiency, and human welfare. Therefore, marine conservation is an essential component of the blue bioeconomy and is crucial for maintaining the health and sustainability of our oceans. Major sectors of the blue bioeconomy are summarized in Figure 1.2.

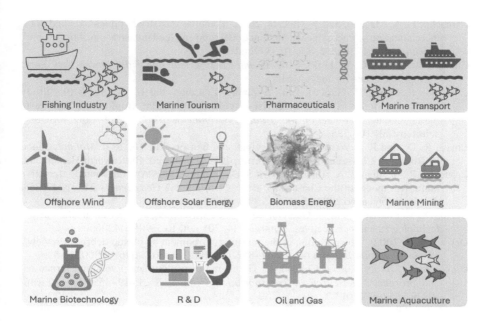

FIGURE 1.2   Important sectors and industries associated with marine bioresources in the blue bioeconomy.

REFERENCES

Adepoju, F. O., M. N. Ivantsova, and O. N. Kanwugu. 2019. "Gray biotechnology: An overview." *AIP Conference Proceedings* 2174 (1). doi: 10.1063/1.5134350

Anderson, C. M., M. J. Krigbaum, M. C. Arostegui, M. L. Feddern, J. Z. Koehn, P. T. Kuriyama, C. Morrisett, C. I. A. Akselrud, M. J. Davis, C. Fiamengo, A. Fuller, Q. Lee, K. N. McElroy, M. Pons, and J. Sanders. 2018. "How commercial fishing effort is managed." *Fish and Fisheries* 20 (2):268–285. doi: 10.1111/faf.12339

Asanka, S. K. K., H.-S. Kim, and Y.-J. Jeon. 2018. "Edible Korean Seaweed: A Source of Functional Compounds." In *Korean Functional Foods*, edited by K.-Y. Park, D. Y. Kwon, K. W. Lee, and S. Park, 359–384. CRC Press.

Bahaj, A. S., and L. E. Myers. 2003. "Fundamentals applicable to the utilisation of marine current turbines for energy production." *Renewable Energy* 28 (14):2205–2211. doi: 10.1016/s0960-1481(03)00103-4

Barcelos, M. C. S., F. B. Lupki, G. A. Campolina, D. L. Nelson, and G. Molina. 2018. "The colors of biotechnology: General overview and developments of white, green and blue areas." *FEMS Microbiology Letters* 365 (21). doi: 10.1093/femsle/fny239

Bray-Miners, J., R. J. Runciman, and G. Monteith. 2012. "Water skiing biomechanics: A study of advanced skiers." *Proceedings of the Institution of Mechanical Engineers, Part P: Journal of Sports Engineering and Technology* 227 (2):137–146. doi: 10.1177/1754337112444688

Bucher, R., and I. Bryden. 2016. "Overcoming the marine energy pre-profit phase: What classifies the game-changing "array-scale success"?" *International Journal of Marine Energy* 13:180–192. doi: 10.1016/j.ijome.2015.05.002

Camacho, F., A. Macedo, and F. Malcata. 2019. "Potential industrial applications and commercialization of microalgae in the functional food and feed industries: A short review." *Marine Drugs* 17 (6):312. doi: 10.3390/md17060312

Charlier, R. H., and C. W. Finkl. 2009. *Ocean Energy: Tide and Tidal Power.* New York: Springer Science+Business Media.

Chen, H., S. Wang, H. Guo, H. Lin, Y. Zhang, Z. Long, and H. Huang. 2019. "Study of marine debris around a tourist city in East China: Implication for waste management." *Science of the Total Environment* 676:278–289. doi: 10.1016/j.scitotenv.2019.04.335

Christ, R. D., and R. L Wernli. 2014. "Navigational Sensors." In *The ROV Manual*, edited by Robert D. Christ and Robert L. Wernli, 453–475. Oxford: Butterworth-Heinemann.

Crouch, J. R., Y. Shen, J. A. Austin, and M. S. Dinniman. 2008. "An educational interactive numerical model of the Chesapeake Bay." *Computers & Geosciences* 34 (3):247–258. doi: 10.1016/j.cageo.2007.03.017

Curto, D., V. Franzitta, and A. Guercio. 2021. "Sea wave energy. A review of the current technologies and perspectives." *Energies* 14 (20). doi: 10.3390/en14206604

Derraik, J. G. 2002. "The pollution of the marine environment by plastic debris: A review." *Marine Pollution Bulletin* 44 (9):842–852. doi: 10.1016/s0025-326x(02)00220-5

Diakomihalis, M. N. 2007. "Chapter 13 Greek maritime tourism: Evolution, structures and prospects." *Research in Transportation Economics* 21:419–455. doi: 10.1016/s0739-8859(07)21013-3

Eisner, T. 1994. "Chemical prospecting: A global imperative." *Proceedings of the American Philosophical Society* 138 (3):385–393.

El-Ramady, H., N. Abdalla, Z. Fawzy, K. Badgar, X. Llanaj, G. Törős, P. Hajdú, Y. Eid, and J. Prokisch. 2022. "Green biotechnology of oyster mushroom (*Pleurotus ostreatus* L.): A sustainable strategy for myco-remediation and bio-fermentation." *Sustainability* 14 (6):3667. doi: 10.3390/su14063667

Esteban, M. D., J. M. Espada, J. M. Ortega, J.-S. López-Gutiérrez, and V. Negro. 2019. "What about marine renewable energies in Spain?" *Journal of Marine Science and Engineering* 7 (8) :249. doi: 10.3390/jmse7080249

Esteban, M., and D. Leary. 2012. "Current developments and future prospects of offshore wind and ocean energy." *Applied Energy* 90 (1):128–136. doi: 10.1016/j.apenergy.2011.06.011

Estes, M., C. Anderson, W. Appeltans, N. Bax, N. Bednaršek, G. Canonico, S. Djavidnia, E. Escobar, P. Fietzek, M. Gregoire, E. Hazen, M. Kavanaugh, F. Lejzerowicz, F. Lombard, P. Miloslavich, K. O. Möller, J. Monk, E. Montes, H. Moustahfid, M. M. C. Muelbert, F. Muller-Karger, L. E. P. Reeves, E. V. Satterthwaite, J. O. Schmidt, A. M. M. Sequeira, W. Turner, and L. V. Weatherdon. 2021. "Enhanced monitoring of life in the sea is a critical component of conservation management and sustainable economic growth." *Marine Policy* 132:104699. doi: 10.1016/j.marpol.2021.104699

European Commission Directorate General for Maritime Affairs and Fisheries. 2022. Realising the potential of the blue bioeconomy: Key messages and calls to action from the European Marine Board. Luxembourg, European Union: Publications Office of the European Union:

FAO. 2018. *The State of World Fisheries and Aquaculture 2018.* Rome, Italy: FAO.

Fauziah, S. H., M. Rizman-Idid, W. Cheah, K. H. Loh, S. Sharma, M. Bordt, T. Praphotjanaporn, A. A. Samah, J. S. B. Sabaruddin, and M. George. 2021. "Marine debris in Malaysia: A review on the pollution intensity and mitigating measures." *Marine Pollution Bulletin* 167:112258. doi: 10.1016/j.marpolbul.2021.112258

Galeb, H. A., E. L. Wilkinson, A. F. Stowell, H. Lin, S. T. Murphy, P. L. Martin-Hirsch, R. L. Mort, A. M. Taylor, and J. G. Hardy. 2021. "Melanins as sustainable resources for advanced biotechnological applications." *Global Challenges* 5 (2):2000102. doi: 10.1002/gch2.202000102

Gartland, K. M., F. Bruschi, M. Dundar, P. B. Gahan, M. V. Magni, and Y. Akbarova. 2013. "Progress towards the 'Golden Age' of biotechnology." *Current Opinion in Biotechnology* 24 (Suppl 1):S6–13. doi: 10.1016/j.copbio.2013.05.011

Gavrilescu, M., and Y. Chisti. 2005. "Biotechnology: A sustainable alternative for chemical industry." *Biotechnology Advances* 23 (7–8):471–499. doi: 10.1016/j.biotechadv. 2005.03.004

Goodfellow, M., and H. P. Fiedler. 2010. "A guide to successful bioprospecting: Informed by actinobacterial systematics." *Antonie Van Leeuwenhoek* 98 (2):119–142. doi: 10.1007/s10482-010-9460-2

Graziano, M., S.-L. Billing, J. O. Kenter, and L. Greenhill. 2017. "A transformational paradigm for marine renewable energy development." *Energy Research & Social Science* 23:136–147. doi: 10.1016/j.erss.2016.10.008

Haefner, B. 2003. "Drugs from the deep: Marine natural products as drug candidates." *Drug Discovery Today* 8 (12):536–544. doi: 10.1016/s1359-6446(03)02713-2

Hall, C. M. 2001. "Trends in ocean and coastal tourism: The end of the last frontier?" *Ocean & Coastal Management* 44 (9–10):601–618. doi: 10.1016/s0964-5691(01)00071-0

Harcourt, F., A. Angeloudis, and M. D. Piggott. 2019. "Utilising the flexible generation potential of tidal range power plants to optimise economic value." *Applied Energy* 237: 873–884. doi: 10.1016/j.apenergy.2018.12.091

Helfer, F., C. Lemckert, and Y. G. Anissimov. 2014. "Osmotic power with pressure retarded osmosis: Theory, performance and trends – A review." *Journal of Membrane Science* 453:337–358. doi: 10.1016/j.memsci.2013.10.053

Hosseini, H., H. M. Al-Jabri, N. R. Moheimani, S. A. Siddiqui, and I. Saadaoui. 2022. "Marine microbial bioprospecting: Exploitation of marine biodiversity towards biotechnological applications – A review." *Journal of Basic Microbiology* 62 (9):1030–1043. doi: 10.1002/jobm.202100504

Hsu, P.-C., L. Centurioni, H.-J. Shao, Q. Zheng, C.-Y. Lu, T.-W. Hsu, and R.-S. Tseng. 2021. "Surface current variations and oceanic fronts in the Southern East China Sea: Drifter experiments, coastal radar applications, and satellite observations." *Journal of Geophysical Research: Oceans* 126 (10):e2021JC017373. doi: 10.1029/2021jc017373

Jain, A., and V. Tailor. 2020. "Emerging Trends of Biotechnology in Marine Bioprospecting: A New Vision." In *Marine Niche: Applications in Pharmaceutical Sciences*, edited by N. M. Nathani, C. Mootapally, I. R. Gadhvi, B. Maitreya and C. G. Joshi, 1–36. Singapore: Springer Singapore.

Jambeck, J. R., R. Geyer, C. Wilcox, T. R. Siegler, M. Perryman, A. Andrady, R. Narayan, and K. L. Law. 2015. "Marine pollution: Plastic waste inputs from land into the ocean." *Science* 347 (6223):768–771. doi: 10.1126/science.1260352

Kafarski, P. 2012. "Rainbow code of biotechnology." *Chemik* 66 (8):814–816.

Karani, P., and P. Failler. 2020. "Comparative coastal and marine tourism, climate change, and the blue economy in African large marine ecosystems." *Environmental Development* 36:100572. doi: 10.1016/j.envdev.2020.100572

Kauling, R. M., R. Rienks, J. A. A. E. Cuypers, H. T. Jorstad, and J. W. Roos-Hesselink. 2023. "SCUBA diving in adult congenital heart disease." *Journal of Cardiovascular Development and Disease* 10 (1):20. doi: 10.3390/jcdd10010020

Kennedy, M. J. 1991. "The evolution of the word 'biotechnology'." *Trends in Biotechnology* 9 (1):218–220. doi: 10.1016/0167-7799(91)90073-q

Kim, L. J., M. Ohashi, Z. Zhang, D. Tan, M. Asay, D. Cascio, J. A. Rodriguez, Y. Tang, and H. M. Nelson. 2021. "Prospecting for natural products by genome mining and microcrystal electron diffraction." *Nature Chemical Biology* 17 (8):872–877. doi: 10.1038/s41589-021-00834-2

Kumar, H. 2020. "Biotechnology: Discoveries and Their Applications in Societal Welfare." In *Biotechnology Business: Concept to Delivery*, edited by A. Saxena, 3–44. Cham: Springer International Publishing.

Kumar, V., R. L. Shrivastava, and S. P. Untawale. 2015. "Solar energy: Review of potential green & clean energy for coastal and offshore applications." *Aquatic Procedia* 4: 473–480. doi: 10.1016/j.aqpro.2015.02.062

Lam, V. W. Y., E. H. Allison, J. D. Bell, J. Blythe, W. W. L. Cheung, T. L. Frölicher, M. A. Gasalla, and U. R. Sumaila. 2020. "Climate change, tropical fisheries and prospects for sustainable development." *Nature Reviews. Earth & Environment* 1 (9):440–454. doi: 10.1038/s43017-020-0071-9

Langer, J., J. Quist, and K. Blok. 2020. "Recent progress in the economics of ocean thermal energy conversion: Critical review and research agenda." *Renewable and Sustainable Energy Reviews* 130:109960. doi: 10.1016/j.rser.2020.109960

Leal, M. C., J. Puga, J. Serodio, N. C. Gomes, and R. Calado. 2012. "Trends in the discovery of new marine natural products from invertebrates over the last two decades: Where and what are we bioprospecting?" *PLoS One* 7 (1):e30580. doi: 10.1371/journal.pone.0030580

Li, C., H. Wang, and P. Sun. 2020. "Numerical investigation of a two-element wingsail for ship auxiliary propulsion." *Journal of Marine Science and Engineering* 8 (5):333. doi: 10.3390/jmse8050333

Liu, W., X. Xu, F. Chen, Y. Liu, S. Li, L. Liu, and Y. Chen. 2020. "A review of research on the closed thermodynamic cycles of ocean thermal energy conversion." *Renewable and Sustainable Energy Reviews* 119:109581. doi: 10.1016/j.rser.2019.109581

Ludvigsen, M., and A. J. Sørensen. 2016. "Towards integrated autonomous underwater operations for ocean mapping and monitoring." *Annual Reviews in Control* 42:145–157. doi: 10.1016/j.arcontrol.2016.09.013

Manikkam, R., P. Pati, S. Thangavel, G. Venugopal, J. Joseph, B. Ramasamy, and S. G. Dastager. 2019. "Distribution and Bioprospecting Potential of Actinobacteria from Indian Mangrove Ecosystems." In *Microbial Diversity in Ecosystem Sustainability and Biotechnological Applications*, edited by T. Satyanarayana, B. N. Johri and S. K. Das, 319–353. Singapore: Springer Singapore.

Martin, R. E. 1990. "A History of the Seafood Industry." In *The Seafood Industry*, edited by R. E. Martin and G. J. Flick, 1–16. Boston, MA: Springer.

Martínez Vázquez, R., J. Milán García, and J. De Pablo Valenciano. 2021. "Analysis and trends of global research on nautical, maritime and marine tourism." *Journal of Marine Science and Engineering* 9 (1):93. doi: 10.3390/jmse9010093

McKendry, P. 2002. "Energy production from biomass (Part 1): Overview of biomass." *Bioresource Technology* 83 (1):37–46. doi: 10.1016/s0960-8524(01)00118-3

McNicholas, G., and M. Cotton. 2019. "Stakeholder perceptions of marine plastic waste management in the United Kingdom." *Ecological Economics* 163:77–87. doi: 10.1016/j.ecolecon.2019.04.022

Miller, R. G., Z. L. Hutchison, A. K. Macleod, M. T. Burrows, E. J. Cook, K. S. Last, and B. Wilson. 2013. "Marine renewable energy development: Assessing the Benthic Footprint at multiple scales." *Frontiers in Ecology and the Environment* 11 (8):433–440. doi: 10.1890/120089

Nachtane, M., M. Tarfaoui, I. Goda, and M. Rouway. 2020. "A review on the technologies, design considerations and numerical models of tidal current turbines." *Renewable Energy* 157:1274–1288. doi: 10.1016/j.renene.2020.04.155

Nguyen, T. T., K. Heimann, and W. Zhang. 2020. "Protein recovery from underutilised marine bioresources for product development with nutraceutical and pharmaceutical bioactivities." *Marine Drugs* 18 (8):391. doi: 10.3390/md18080391

Orlandi, V. T., E. Martegani, C. Giaroni, A. Baj, and F. Bolognese. 2022. "Bacterial pigments: A colorful palette reservoir for biotechnological applications." *Biotechnology and Applied Biochemistry* 69 (3):981–1001. doi: 10.1002/bab.2170

Oyemitan, I. A. 2017. "African Medicinal Spices of Genus Piper." In *Medicinal Spices and Vegetables from Africa*, edited by V. Kuete, 581–597. Academic Press.

PADI. 2021. "Padi Global Statistics 2016–2021." Accessed November 5. https://www.padi. com/sites/default/files/documents/2022-08/ABOUT%20PADI%20-%20Global%20 Statistics%20%20%2716-%2721.pptx%20%281%29.pdf

Paiva, S. V., P. B. M. Carneiro, T. M. Garcia, T. C. L. Tavares, L. de Souza Pinheiro, A. R. X. Neto, T. C. Montalverne, and M. O. Soares. 2023. "Marine carbonate mining in the Southwestern Atlantic: Current status, potential impacts, and conservation actions." *Marine Policy* 148:105435. https://doi.org/10.1016/j.marpol.2022.105435

Paolo, F., A. Viola, M. Carta, D. Secci, G. Fancello, and P. Serra. 2020. "Treatment of Port Wastes according to the Paradigm of the Circular Economy." International Conference on Computational Science and Its Applications (ICCSA 2020), Cham.

Papageorgiou, M. 2016. "Coastal and marine tourism: A challenging factor in marine spatial planning." *Ocean & Coastal Management* 129:44–48. doi: 10.1016/j.ocecoaman.2016.05.006

Prabhakaran, S., V. Nair, V. Nair, and S. Ramachandran. 2013. "Marine waste management indicators in a tourism environment." *Worldwide Hospitality and Tourism Themes* 5 (4):365–376. doi: 10.1108/whatt-03-2013-0013

Roberts, L. 1992. "Chemical prospecting: Hope for vanishing ecosystems?" *Science* 256 (5060):1142–1143. doi: 10.1126/science.256.5060.1142

Rocha, J., L. Peixe, M. Gomes, and R. Calado. 2011. "Cnidarians as a source of new marine bioactive compounds: An overview of the last decade and future steps for bioprospecting." *Marine Drugs* 9 (10):1860–1886. doi: 10.3390/md9101860

Rona, Peter A. 2008. "The changing vision of marine minerals." *Ore Geology Reviews* 33 (3–4):618–666. doi: 10.1016/j.oregeorev.2007.03.006

Rotter, A., et al. 2021. "The essentials of marine biotechnology." *Frontiers in Marine Science* 8. doi: 10.3389/fmars.2021.629629

Rourke, F. O., F. Boyle, and A. Reynolds. 2010. "Marine current energy devices: Current status and possible future applications in Ireland." *Renewable and Sustainable Energy Reviews* 14 (3):1026–1036. doi: 10.1016/j.rser.2009.11.012

Runciman, R. J. 2011. "Water-skiing biomechanics: A study of intermediate skiers." *Proceedings of the Institution of Mechanical Engineers, Part P: Journal of Sports Engineering and Technology* 225 (4):231–239. doi: 10.1177/1754337111403693

Sala, E., et al. 2021. "Protecting the global ocean for biodiversity, food and climate." *Nature* 592 (7854):397–402. doi: 10.1038/s41586-021-03371-z

Sanjeewa, K. K. A., H.-S. Kim, H.-G. Lee, T. U. Jayawardena, D. P. Nagahawatta, H.-W. Yang, D. Udayanga, J.-I. Kim, and Y.-J. Jeon. 2021. "3-Hydroxy-5,6-epoxy-β-ionone isolated from invasive harmful brown seaweed *Sargassum horneri* protects MH-S mouse lung cells from urban particulate matter-induced inflammation." *Applied Sciences* 11 (22). doi: 10.3390/app112210929

Sbragaglia, V., R. Arlinghaus, D. T. Blumstein, H. Diogo, V. J. Giglio, A. Gordoa, F. A. Januchowski-Hartley, M. Laporta, S. J. Lindfield, J. Lloret, B. Mann, D. McPhee, J. A. C. C. Nunes, P. Pita, M. Rangel, O. K. Rhoades, L. A. Venerus, and S. Villasante. 2023. "A global review of marine recreational spearfishing." *Reviews in Fish Biology and Fisheries* 33 (4):1199–1222. doi: 10.1007/s11160-023-09790-7

Shields, M. A., D. K. Woolf, M. Grist, S. A. Kerr, A. C. Jackson, R. E. Harris, M. C. Bell, R. Beharie, A. Want, E. Osalusi, S. W. Gibb, and J. Side. 2011. "Marine renewable energy: The ecological implications of altering the hydrodynamics of the marine environment." *Ocean & Coastal Management* 54 (1):2–9. doi: 10.1016/j.ocecoaman.2010.10.036

Strobel, G., and B. Daisy. 2003. "Bioprospecting for microbial endophytes and their natural products." *Microbiology and Molecular Biology Reviews: MMBR* 67 (4):491–502. doi: 10.1128/MMBR.67.4.491-502.2003

Sumaila, U. R., C. Bellmann, and A. Tipping. 2016. "Fishing for the future: An overview of challenges and opportunities." *Marine Policy* 69:173–180. doi: 10.1016/j.marpol.2016.01.003

Teague, J., M. J. Allen, and T. B. Scott. 2018. "The potential of low-cost ROV for use in deep-sea mineral, ore prospecting and monitoring." *Ocean Engineering* 147:333–339. doi: 10.1016/j.oceaneng.2017.10.046

Tiwari, S., and P. S. Chauhan. 2021. "Ecological Restoration and Plant Biodiversity." In *Bioprospecting of Plant Biodiversity for Industrial Molecules*, edited by S.K. Upadhyay and S. P. Singh, 91–97. New York: John Wiley & Sons, Inc.

Tsai, F. M., T.-D. Bui, M.-L. Tseng, M. K. Lim, and R. R. Tan. 2021. "Sustainable solid-waste management in coastal and marine tourism cities in Vietnam: A hierarchical-level approach." *Resources, Conservation and Recycling* 168:105266. doi: 10.1016/j.resconrec.2020.105266

Vega, L. A. 2002. "Ocean thermal energy conversion primer." *Marine Technology Society Journal* 36 (4):25–35. doi: 10.4031/002533202787908626

Vieira, H., M. C. Leal, and R. Calado. 2020. "Fifty shades of blue: How blue biotechnology is shaping the bioeconomy." *Trends in Biotechnology* 38 (9):940–943. doi: 10.1016/j.tibtech.2020.03.011

Wang, Z., R. Carriveau, D. S.-K. Ting, W. Xiong, and Z. Wang. 2019. "A review of marine renewable energy storage." *International Journal of Energy Research* 43 (12): 6108–6150. doi: 10.1002/er.4444

Waters, S., and G. Aggidis. 2016. "Tidal range technologies and state of the art in review." *Renewable and Sustainable Energy Reviews* 59:514–529. doi: 10.1016/j.rser.2015.12.347

Wei, N., J. Quarterman, and Y. S. Jin. 2013. "Marine macroalgae: An untapped resource for producing fuels and chemicals." *Trends in Biotechnology* 31 (2):70–77. doi: 10.1016/j.tibtech.2012.10.009

Wijesekara, I., and S.-K. Kim. 2015. "Application of Marine Algae Derived Nutraceuticals in the Food Industry." In *Marine Algae Extracts*, edited by S.-K. Kim and K. Chojnacka, 627–638. Hoboken, New Jersey: Wiley-VCH Verlag GmbH & Co. KGaA.

Wilks, J. 2021. "Safety in Coastal and Marine Tourism." In *Tourist Health, Safety and Wellbeing in the New Normal*, edited by J. Wilks, D. Pendergast, P. A. Leggat and D. Morgan, 411–442. Singapore: Springer Singapore.

Xin, B., C. Zhong, and Y. Wang. 2023. "Integrating the marine carbon resource manni-tol into biomanufacturing." *Trends in Biotechnology* 41 (6):745–749. doi: 10.1016/j.tibtech.2022.12.010

Yang, X., N. Liu, P. Zhang, Z. Guo, C. Ma, P. Hu, and X. Zhang. 2019. "The current state of marine renewable energy policy in China." *Marine Policy* 100:334–341. doi: 10.1016/j.marpol.2018.11.038

Yellen, J. E., A. S. Brooks, E. Cornelissen, M. J. Mehlman, and K. Stewart. 1995. "A middle Stone Age worked bone industry from Katanda, Upper Semliki Valley, Zaire." *Science* 268 (5210):553–556. doi: 10.1126/science.7725100

Yeung, A. W. K., N. T., et al. 2019. "Current research in biotechnology: Exploring the biotech forefront." *Current Research in Biotechnology* 1:34–40. doi: 10.1016/j.crbiot.2019.08.003

Young, M. A., S. Foale, and D. R. Bellwood. 2015. "Dynamic catch trends in the history of recreational spearfishing in Australia." *Conservation Biology* 29 (3):784–794. doi: 10.1111/cobi.12456

Zhang, Y.-l., Z. Lin, and Q.-l. Liu. 2014. "Marine renewable energy in China: Current status and perspectives." *Water Science and Engineering* 7 (3):288–305.

2 Seaweeds

2.1 GENERAL FACTS ABOUT SEAWEEDS

Seaweeds, also known as marine macroalgae, are a diverse group of multicellular, photosynthetic organisms found in marine environments. Usually, the term 'seaweed' is used to describe macroscopic, multicellular marine algal species. Seaweeds have been used as human food since ancient times (Mahadevan 2015). With the high market demand and nutritional properties associated with seaweeds, the cultivation of seaweed is among the rapidly expanding industries worldwide. Its farming regions cover an extensive area of 48 million km², distributed across 132 countries, with 37–44 of these nations actively engaged in seaweed production (Froehlich et al. 2019). According to recent reports, global seaweed production has increased from 0.65 million tons in 1950 to 35.82 million tons in 2019, predominantly through aquaculture practices, which account for 97% of the total output (Cai et al. 2021; Park et al. 2023). The seaweed industry has a thriving worldwide market valued at approximately US$11.8 billion (Sultana et al. 2023). In East Asian nations, people who lived close to the seaside usually consume seaweed as a soup, main meal, or side dish (Gomez-Gutierrez et al. 2011). Usually, Europeans consume less amount of seaweed compared to Asians due to the regulations and the food habits of Europeans (Syakilla et al. 2022). Nonetheless, during the past several decades, the consumption of seaweed has steadily expanded in Europe due to the discovery of potential health benefits associated with seaweeds (Asanka, Kim, and Jeon 2018).

2.2 BOTANICAL FACTS ABOUT SEAWEEDS

The ecology of seaweeds is dominated by two major environmental conditions: light, which is essential for photosynthesis, and marine environment to establish habitats. In addition to the above major requirements, a firm attachment point is also essential for seaweed growth. Therefore, seaweed habitats only report from the intertidal zones and subtidal zones until light is available for their photosynthesis (Nautiyal 2013). Seaweeds are an essential part of marine ecosystems, providing food and habitat for numerous marine species, and playing a critical role in maintaining oceanic nutrient cycles (Harley et al. 2012). Seaweeds have a unique structure that enables them to survive in harsh marine environments. Seaweeds have no true roots, but instead have holdfasts that attach them to rocks or other surfaces. Seaweeds have a blade-like structure, which is the main photosynthetic organ that captures sunlight for energy production. They also have a stipe, which is a stem-like structure that supports the blades and helps transport nutrients and water. Some seaweeds also have air bladders that help them float

DOI: 10.1201/9781003477365-2

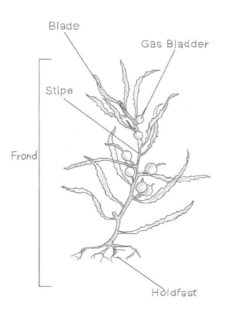
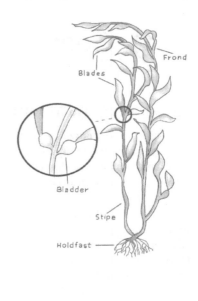

FIGURE 2.1   Structure of a typical seaweed.

and remain near the surface for optimal light exposure (Figure 2.1) (Sudhakar et al. 2018). Seaweeds have a high level of structural complexity due to their ability to form multicellular tissues and organs. They have a well-developed system of internal structures, including vascular systems that help transport nutrients and water, and specialized cells that allow for gas exchange and nutrient uptake (Chapman 2012).

Holdfast: It is a unique root-like structure that holds the plant to the bottom surface or rocks. It does not penetrate through the sand or mud nor aid in gathering nutrients.

Stipe: Stipe is a stem-like structure found in some macroalgae species. The stipe can be long and tough to provide support. It acts as an attachment for the blades. It carries sugars from the blades to the rest of the plant.

Floats: Gas-filled bladders or floats are an organ between the blade and stipe. Floats may contain carbon monoxide (CO) or pneumatocysts. The action of float helps to keep the blades near the water surface and this action is useful to capture more sunlight for photosynthesis.

Frond: It is referred to as the combined part of the blade and stipe.

Lamina or Blades: The main photosynthetic leaf-like flattened portions and blades are useful to absorb sunlight.

Thallus: The simple or more advanced body-like structure is known as a thallus. All sections of the thallus can photosynthesize.

2.3 REPRODUCTION IN SEAWEEDS

In general, seaweeds exhibit both asexual and sexual reproduction mechanisms. Asexual reproduction in seaweeds does not entail fertilization, whereas sexual reproduction involves the fusion of gametes (Charrier et al. 2017). One common method of vegetative reproduction in seaweeds is the formation of zoospores. Within the Chlorophyceae group, various seaweeds employ all three sexual reproduction modes: isogamous, anisogamous, and oogamous (Baweja et al. 2016). However, the growth and reproduction of seaweeds depend on environmental factors, such as temperature of the water column, light intensity, season, wave motion and mechanical shock, nutrient availability, pH, osmotic stress, and biotic factors such as growth stage of seaweed, bacterial association, extracellular products, and predators (Agrawal 2012; Navarro, Huovinen, and Gómez 2020). Typically, most documented seaweed species are capable of both asexual and sexual reproduction (Baweja et al. 2016).

2.3.1 Asexual Reproduction

Asexual reproduction in seaweeds is a process by which new seaweeds are generated without the involvement of fertilization or the fusion of gametes. This type of reproduction allows seaweeds to produce genetically identical generations continuously (Hiraoka et al. 2003; Moreira et al. 2021). Several types of asexual reproduction methods are reported from seaweeds, including thalli fragmentation, propagules, and spores (de Bettignies et al. 2018; Hoffmann 1987; Leal, Hurd, and Roleda 2014; Moreira et al. 2021). In the following sections, each of them is explained briefly,

2.3.1.1 Thalli Fragmentation

Seaweeds can undergo asexual reproduction by fragmenting their thalli. In this process, each fragmented piece has the ability to generate a new thallus and reconstitute a distinct individual. This method of reproduction contributes to the proliferation and regeneration of seaweed populations (Otaíza et al. 2018).

2.3.1.2 Propagules

Seaweeds can reproduce asexually through propagules, which are small clusters of cells. These propagules have the capacity to attach to a substrate and initiate the development of a new thallus, or the main body of the seaweed. This process allows for the efficient expansion and colonization of seaweed populations (de Bettignies, Wernberg, and Gurgel 2018; Clayton 1992).

2.3.1.3 Spores

Seaweeds can asexually reproduce through the formation and release of spores (Hoffmann 1987). Spores originate within sporocysts, which develop from modified mother cells through mitotic divisions of the mother cell's nucleus. Once the spores have completed their differentiation, they are released from the sporocysts into the outside through an opening in their protective wall.

The mobility of spores varies among species. Some spores are immobile, known as aplanospores, and they float within the water column. Others are mobile, referred to as zoospores, and they move by means of whip-like flagella (Anderson 2007; Zechman and Mathieson 1985). Regardless of their mobility, the primary purpose of spores is to promote the dispersal of the species. Spores attach themselves to new and distant substrates, germinate, and eventually develop into new individuals.

2.3.2 Sexual Reproduction

Sexual reproduction in seaweeds involves the production and fusion of gametes. In green and brown seaweeds, these gametes are spermatozoids and ovum (Callow 1985). In red seaweeds, spermatia (sperm) and carpogonium (egg) play similar roles (Mayanglambam and Sahoo 2015; Shim et al. 2020). This process results in the formation of zygotes.

Zygotes, also known as thick-wall zygospores, are typically formed in terrestrial and freshwater algae. They serve as resilient structures capable of enduring stressful conditions, such as water scarcity and temperature extremes. However, the advantages of zygospore formation are balanced by the substantial energy required to create and maintain these cells in a near-dormant state over extended periods. Additionally, zygospores may miss out on numerous potential cell divisions that could have occurred during the resting phase. In contrast, seaweeds, such as red and brown, typically produce thin-walled zygotes that are not dormant. These zygotes promptly divide after their formation. The continuous presence of water in the marine environment negates the need for the protective features found in zygospores of terrestrial and freshwater algae (Agrawal 2012).

In some seaweeds, the gametes of different sexes exhibit similar morphology, resulting in a reproductive process known as isogamous. In contrast, in some seaweeds, male and female gametes have different morphologies, leading to a heterogamous mode of reproduction. In both cases, following the fusion of these gametes, a cell known as either an 'egg' or a 'zygote' (with a chromosome count of $2n$) is formed (Figure 2.2) (Pereira 2021).

2.4 CLASSIFICATION OF SEAWEEDS

Based on the pigments, seaweeds can be classified into three major groups: brown (phaeophyta), green (chlorophyta), and red seaweeds (rhodophyta) (Peng et al. 2015). Brown seaweeds produce fucoxanthin, a brown color xanthophyll pigment that masks other pigments such as chlorophyll a and chlorophyll c, giving them a brown appearance. Green seaweeds produce chlorophyll a and chlorophyll b in similar quantities to higher plants, giving them a green appearance. Red seaweeds produce phycoerythrin and phycocyanin, which masks other green and yellow color pigments, including chlorophyll and beta-carotene, giving them a red appearance (Corino et al. 2019; Gupta and Abu-Ghannam 2011; Peng et al. 2015).

2.5 COMPOSITION OF SEAWEEDS

When considering the composition of seaweeds, fresh seaweed biomass contains a high amount of moisture and can account for up to 94% of the biomass. Like

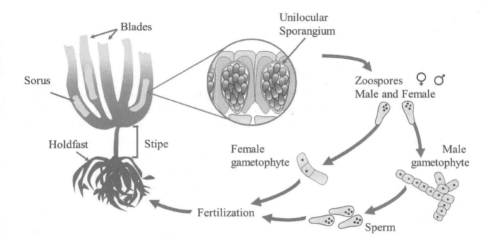

FIGURE 2.2 General life cycle of seaweeds.

other plants, seaweeds contain proteins, lipids, carbohydrates, vitamins such as A, B1/2/9/12, C, D, and E), and minerals (Ca, Fe, I, Mg, P, K, Zn, Cu, Mn, Se, and F) (Park et al. 2023). However, some studies revealed that the nutrient composition of seaweeds depends on the growing season and the area of production (Park et al. 2023). Traditionally, seaweeds have been utilized as cuisines and medicines in Asian countries such as Korea, China, and Japan (Jeon and Sanjeewa 2018b). In addition to the nutritional value of seaweeds, there are accumulated evidence suggesting that the metabolites (polysaccharides, phlorotannins, lipids, and sterols) synthesized in seaweeds reported to have interesting bioactive properties such as anticancer, anti-inflammation, and antioxidant (Holdt and Kraan 2011). Thus, food products containing seaweed as a functional ingredient appear to be continually growing in the market. Moreover, recent studies have demonstrated that polysaccharides, polyphenols, carotenoids, and sterols separated from brown seaweeds have the potential to develop as functional materials due to their promising bioactivities.

2.6 BIOACTIVE COMPOUNDS REPORTED FROM EDIBLE SEAWEEDS

As we discussed in Section 2.1, seaweeds have gained significant attention in recent years due to their potential health benefits. These seaweeds are known to be rich sources of bioactive compounds such as fucoidan, laminarin, alginic acid, fucoxanthin, carrageenan, ulvan, phlorotannins, proteins, pigments, and some bioactive sterols (Park et al. 2023; Sanjeewa et al. 2017; Syakilla et al. 2022). These compounds have been shown to have various beneficial effects on human health, including anti-inflammatory, antioxidant, antitumor, anticoagulant, antiviral, antibacterial, immunomodulatory, cardioprotective, and neuroprotective properties (Asanka, Kim, and Jeon 2018; Fernando et al. 2018; Holdt and Kraan 2011; Park et al. 2023; Sanjeewa et al. 2017; Syakilla et al. 2022; Jeon and Sanjeewa 2018a). In this context, understanding the bioactive compounds present in edible brown seaweeds is essential for exploring their potential use as functional foods and nutraceuticals.

2.6.1 Seaweed Polysaccharides

Polysaccharides found in seaweeds possess a variety of bioactive properties that can be used in several medical, pharmaceutical, and biotechnological applications. These applications include the development of anticoagulant, anti-inflammatory, antioxidant, immunostimulatory, and antitumor agents. Therefore, these macromolecules have gained interest due to their potential usefulness in multiple fields (Groth, Grünewald, and Alban 2009). Several studies have demonstrated that polysaccharides derived from marine sources, specifically seaweeds, have superior safety profiles compared to those derived from mammals. This feature makes them particularly attractive for applications in drug discovery and other industries, where safety is a critical concern. Moreover, the ability to produce seaweed polysaccharides in large quantities at a low cost is another significant advantage. These characteristics make seaweed polysaccharides a promising source for the development of novel therapeutics and other industrial applications (Blondin and de Agostiniz 1995; Senni et al. 2011).

2.6.1.1 Fucoidans

Fucoidans are a class of sulfated polysaccharides found in the cell walls of brown seaweeds, and they have attracted significant interest due to their numerous bioactive properties (Sanjeewa et al. 2019). These bioactive properties include antioxidant, anticancer, anticoagulant, and immune modulatory effects, which make them promising candidates for potential therapeutic applications (Synytsya et al. 2010).

Fucoidans are composed of a variety of monosaccharides, including L-fucose, galactose, sulfate, guluronic acid, rhamnose, arabinose, and xylose. L-Fucose is the most abundant monosaccharide in fucoidans, with the sulfation pattern and position varying depending on the source of the seaweed. The primary structure of fucoidans consists of a backbone of $(1 \rightarrow 3)$-linked α- L-fucose, with a repeating sequence of alternating $\alpha(1 \rightarrow 3)$ and $\alpha(1 \rightarrow 4)$ glycosidic bonds (Figure 2.3). However, the structural diversity of fucoidans is quite significant, with significant differences in sulfation patterns and branching, which can result in significant structural variability (Berteau and Mulloy 2003). This structural variability of fucoidans is closely related to their bioactivity. For example, some studies have shown that fucoidans with higher sulfate content exhibit greater anticoagulant and anti-inflammatory activity. Similarly, fucoidans with a higher degree of branching have been shown to have a stronger immune modulatory effect. Therefore, fucoidans exhibit significant structural and functional diversity, which makes them an exciting area of research for the development of novel therapeutic agents (Sanjeewa et al. 2021).

2.6.1.2 Laminarin

Laminarin is the primary storage of polysaccharide found in brown seaweeds, accounting for approximately 35% of the algal dry weight (Figure 2.4) (Li et al. 2022; Makkar et al. 2016). It is a linear polysaccharide that consists of $(1,3)$-β-D-glucan

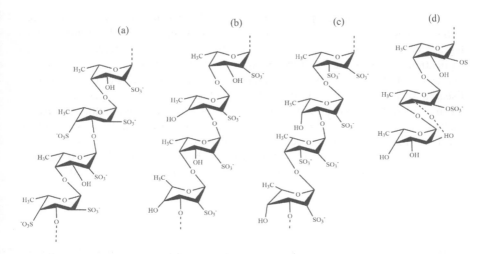

FIGURE 2.3 Structural motifs of fucoidans separated from different brown seaweeds.

with β-(1,6) branching (Smith et al. 2018). The structure of laminarin can be divided into two groups based on the reducing end: M and G chains. M chains terminate with a mannitol residue, while G chains end with a glucose residue. The ratio of M to G chains in laminarin varies between different species of brown seaweeds and can also change depending on environmental factors such as light and temperature (Rioux, Turgeon, and Beaulieu 2007a).

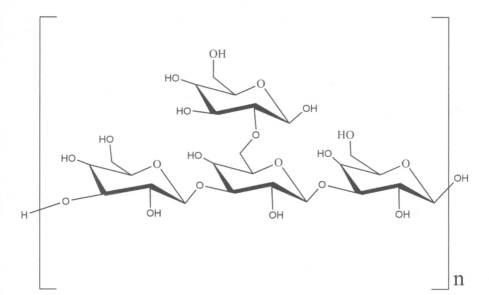

FIGURE 2.4 General structure of laminarin, a storage polysaccharide synthesized in brown seaweeds.

Laminarin is an underexploited polysaccharide reported from brown seaweeds with strong potential in therapeutic and nutraceutical applications. Laminarin is biodegradable, biocompatible, and nontoxic, making it a promising material for various applications (Makkar et al. 2016). Studies have shown that laminarin has therapeutic potential in promoting and enhancing health and may help protect against diseases such as cardiovascular diseases, metabolic disorders, cancer, diabetes, obesity, anti-inflammatory activity, osteoarthritis, oral diseases, multiple sclerosis, Alzheimer's and Parkinson's diseases, and vision-improving agents (Smith et al. 2018). Furthermore, laminarin acts as a modulator of intestinal metabolism by influencing intestinal pH and the production of short-chain fatty acids. In addition, laminarins are also an excellent source of dietary fiber (Karuppusamy et al. 2022).

2.6.1.3 Alginic Acid

Alginate is a complex polysaccharide that serves as a major component of the cell walls of brown seaweeds and the intercellular matrix (Andrade et al. 2004). It is a vital structural polysaccharide that imparts strength and rigidity to the algal cell walls, enabling them to resist the pressure of the surrounding water (Chen, Yin, and Wang 2022). Alginic acids are alkali-soluble polysaccharides, while other cell wall constituents such as fucoidan and laminarin are water-soluble. Seaweed typically contains 10–40% alginic acid by dry weight, with the exact amount heavily dependent on the depth at which the seaweed grows and the season in which it is cultivated (Rioux, Turgeon, and Beaulieu 2007a). In general, the structure of alginic acid is composed of linear polysaccharides containing 1,4-linked β-D-mannuronic and α-l-guluronic acid residues arranged simultaneously in blockwise order along the chain (Figure 2.5). The molecular weight of alginate typically falls between 500 and 1,000 kDa (Rioux, Turgeon, and Beaulieu 2007a; Rioux, Turgeon, and Beaulieu 2007b). Its solubility is affected by various factors, such as pH, concentration, the presence of ions in solution, the presence of divalent ions, and ionic strength. In the presence of divalent ions like calcium, alginate has a tendency to gel or form a gel-like substance (Rioux, Turgeon, and Beaulieu 2007a).

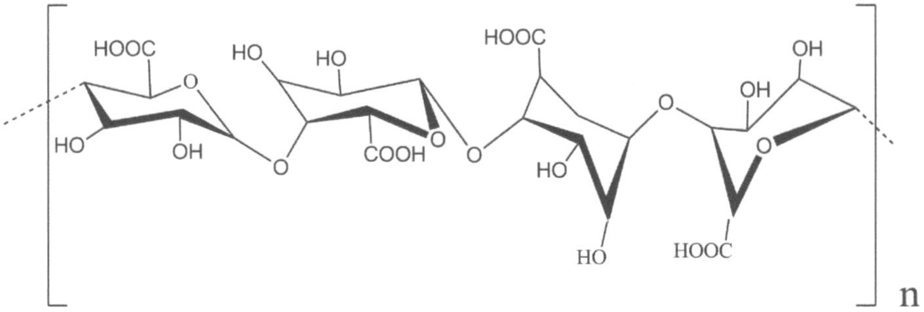

FIGURE 2.5 The chemical structure of the repeating unit's alginic acids.

The pharmaceutical industry widely utilizes brown seaweed alginate in the formulation of various medicines. Alginate has been reported to have beneficial properties such as antioxidant, anti-inflammatory, anti-cancer, and antiviral activities. Due to these properties, alginate is used as a component in the production of several pharmaceuticals such as tablets, capsules, and suspensions (Gupta and Abu-Ghannam 2011). In the food industry, alginate is a popular thickening agent used in the preparation of jellies, beverages, and ice cream. Alginate works by forming a gel-like substance when it comes into contact with calcium ions. This property of alginate makes it an effective thickening agent for various food products (Gupta and Abu-Ghannam 2011; Rashedy et al. 2021).

Alginate is also used in the cosmetic industry as a water-binding and thickening agent. Alginate has the ability to absorb large amounts of water, making it a useful additive in cosmetic formulations. Alginates extracted from seaweeds are commonly found in moisturizers, face masks, and other skincare products (Rashedy et al. 2021). Alginate's water-absorbing capacity also helps to enhance the skin's hydration and improve the overall appearance of the skin. Overall, alginate is a versatile substance that has a wide range of applications in various industries, including pharmaceuticals, food, and cosmetics. Its unique properties make it a valuable ingredient in many products and formulations.

2.6.1.4 Carrageenan

Carrageenans are a group of linear sulfated polysaccharides that are extracted from red seaweeds, primarily *Chondrus crispus* (also known as Irish moss) and other closely related species (Lipinska et al. 2020). Carrageenans are composed of D-galactose units linked alternately with α-1,4 and β-1,3 linkages. These sulfated galactans are classified according to the presence of 3,6-anhydrogalactose on the 4-linked residue and the number and position of the sulfate groups (Figure 2.6) (Relleve et al. 2005). There are three main types of carrageenans: kappa, iota, and lambda (Lipinska et al. 2020; Relleve et al. 2005). Kappa-carrageenan has one sulfate group per disaccharide unit and forms strong gels in the presence of potassium or calcium ions. Iota-carrageenan also has one sulfate group per disaccharide unit, but it forms weaker gels in the presence of calcium ions. Lambda-carrageenan has three sulfate groups per disaccharide unit and does not form gels, but it is widely used as a thickening agent in food and other applications (Zia et al. 2017).

The degree of sulfation and the distribution of sulfate groups along the carrageenan chain affect its properties, including its solubility, gel-forming ability, and viscosity (Cheng et al. 2021). Carrageenans are widely used in the food industry as gelling agents, stabilizers, and thickening agents in a variety of applications, including dairy products, meat products, and processed foods. They are also used in pharmaceuticals, cosmetics, and other industries.

2.6.1.5 Ulvan

The name 'ulvan' was primarily introduced by Lahaye and Axelos to designate the sulfated rhamno-glucuronans obtained from *Ulva* spp. It was derived from the terms 'ulvin' and 'ulvacin' presented by Kylin, which referred to diverse

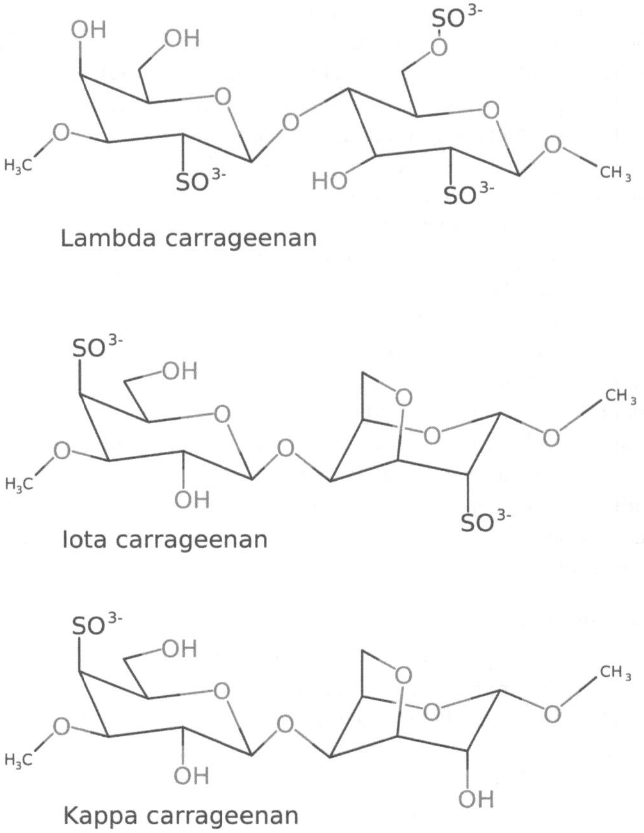

FIGURE 2.6 Repeating unit of the red seaweed polysaccharide, carrageenans.

fractions of *Ulva lactuca* polysaccharides (Robic et al. 2009a). Ulvan, a type of sulfated polysaccharide, is found mostly in green seaweeds and makes up about 9–36% of its dry weight. It is typically sourced from species such as *Ulva, Gayralia,* and *Monostroma.* The backbone of ulvan consists of α- and β-(1→4) linked sugars with repeating disaccharide units that serve as distinctive markers. The two types of ulvanobiuronic acid, A3S, and B3S, differ in their linkage: A3S contains β-D-glucuronic acid linked to (1→4) α-L-rhamnose 3-sulfate, while B3S features α-L-iduronic acid attached to α-L-rhamnose 3-sulfate through a (1→4) linkage. The A3S type of ulvan is more commonly found than other structural types (Glasson et al. 2022). Ulvanobiose U3S is composed of β-D-xylose 2-sulfate linked to α-L-rhamnose 3-sulfate through a (1→4) linkage. Meanwhile, the U2′S,3S type features β-D-xylose attached to α-L-rhamnose 3-sulfate through a (1→4) linkage (Glasson et al. 2022; Robic et al. 2009b; Robic, Sassi, and Lahaye 2008). The major repeating disaccharide units of ulvan are depicted in Figure 2.7.

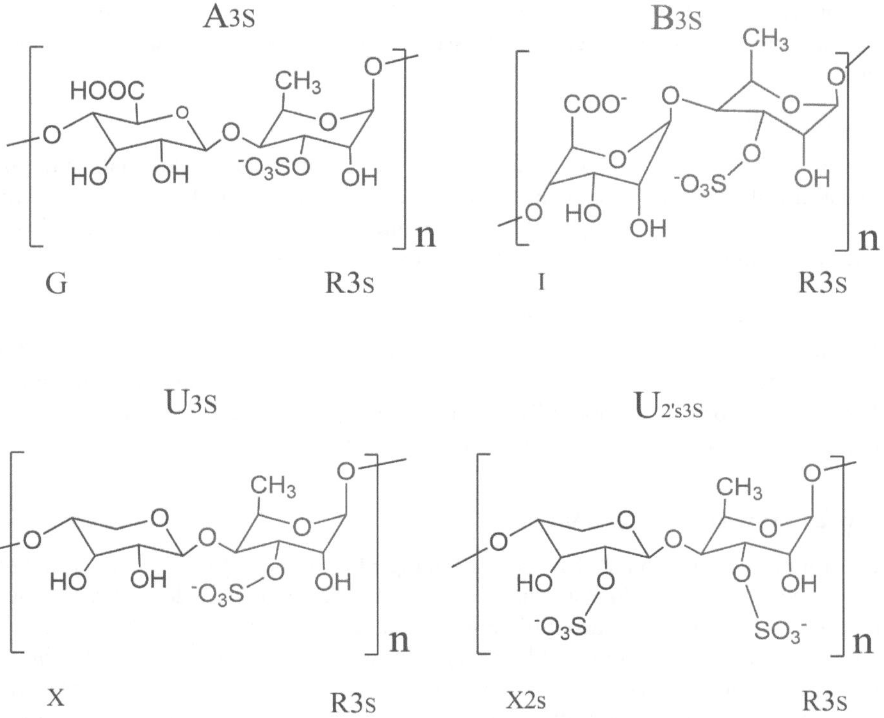

FIGURE 2.7 The main repeating disaccharide units found in various forms of ulvans.

Ulvan holds significant potential for a wide range of industries, including the food industry, agriculture, biomedical applications, cosmetics, and environmental applications. The summary of each application is provided in the following sections.

2.6.1.5.1 Food Industry

Ulvan is a promising ingredient in the food industry, as it has a variety of functional properties. For example, it can be used as a thickener, stabilizer, and emulsifier in food products such as sauces, dressings, and desserts. Additionally, it has antioxidant properties and may be used as a natural preservative in food products.

2.6.1.5.2 Agriculture

Ulvan has been shown to have potential as a plant growth promoter and a natural pesticide. It can enhance the growth of plants and increase their resistance to pests and diseases. Ulvan can also be used as a soil amendment to improve soil structure and nutrient content.

2.6.1.5.3 Biomedical Applications

Ulvan has demonstrated promising potential in diverse biomedical applications such as drug delivery and tissue engineering. Its ability to form gels and nanoparticles makes it a promising material for drug delivery. Additionally, ulvan can support the growth of cells and tissues, making it a useful material for tissue engineering.

2.6.1.5.4 Cosmetics

Ulvan has been shown to have potential in the cosmetic industry due to its antiaging and anti-inflammatory properties. It can be used in skincare products such as creams and serums to reduce the appearance of wrinkles and fine lines. Ulvan can also be used as a natural alternative to synthetic preservatives in cosmetics.

2.6.1.5.5 Environmental Applications

Ulvan has potential applications in the environmental industry as a natural adsorbent and bioremediation agent. It can be used to remove heavy metals and other pollutants from water and soil. Additionally, ulvan can be used as a fertilizer for plants, as it contains nutrients such as nitrogen and phosphorus.

2.6.2 SEAWEED PHLOROTANNIN

Phlorotannins are produced by polymerizing phloroglucinol (1,3,5-trihydroxyben-zene) monomer units through the acetate-malonate or polyketide pathway. These compounds are highly hydrophilic and have a broad molecular weight range, typically between 126 Da and 650 kDa. Marine brown algae can accumulate various types of phlorotannins, with low, intermediate, and high molecular weights that contain both phenyl and phenoxy units. The chemical structures of phenols can be defined as aromatic rings of carbon that bear one or more hydroxyl substituents. According to Handique and Baruah (2002) model, polyphenol compounds are capable of pairing or reacting with one-electron oxidants formed during biological processes. As a result of this mode of action, polyphenol compounds are believed to be the major bioactive compounds found in biological systems. Phlorotannins isolated from brown algae represent the most exclusively studied class of marine secondary metabolites. Phlorotannins are considered as important naturally occurring secondary metabolites with potential uses in economically important industries such as pharmacology and the functional food industry (Mohamed, Hashim, and Rahman 2012; Xiaojun et al. 1996). According to previous studies, phlorotannins have a wide range of therapeutic biological actions, including antimicrobial, antioxidant, anti-adipogenesis, antidiabetic, anticancer, immunomodulatory, anti-inflammatory, anti-obesity, and numerous other biomedical applications (Khan et al. 2022). Furthermore, the wide range of biological activities attributed to phlorotannins is expected to enhance their beneficial health properties in the food, pharmaceutical, and cosmeceutical industries. Despite the promising potential of phlorotannins, the successful creation of a range of nutraceutical products derived from brown algal polyphenols is yet to be achieved.

Figure 2.8 illustrates the chemical structures of some phlorotannins isolated from brown seaweeds, including the commonly studied phloroglucinol, eckol, dioxinodehydroeckol, fucodiphlorethol-G, and dieckol. Depending on their linkage, phlorotannins can be classified into four subclasses: fuhalols and phlorethols

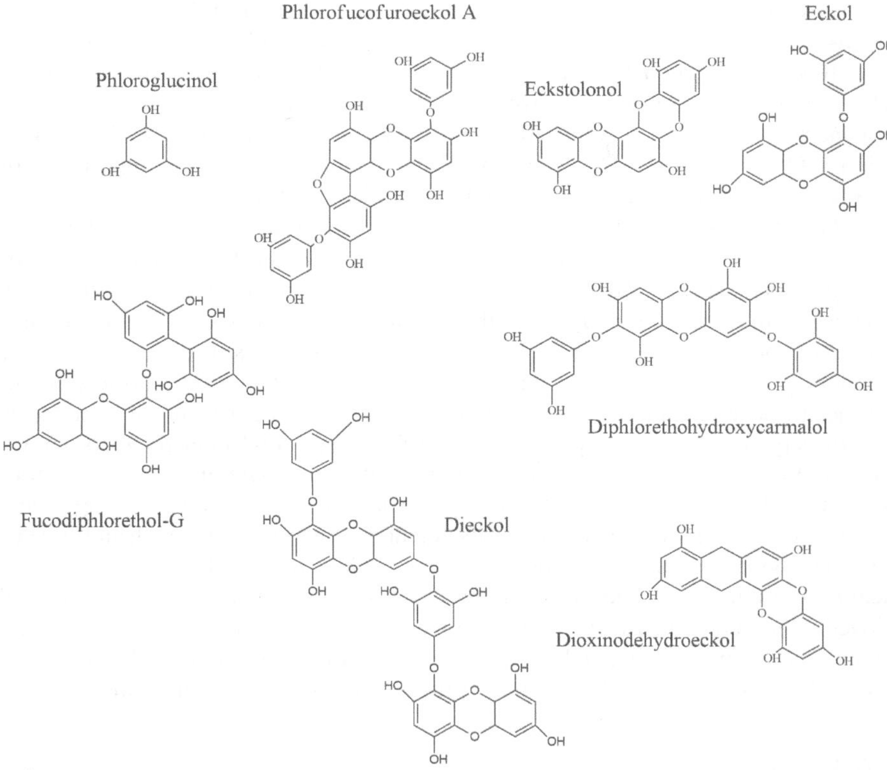

FIGURE 2.8 Bioactive phlorotannin reported from brown seaweeds.

(with an ether linkage), fucols (with a phenyl linkage), fucophloroethols (with both an ether and phenyl linkage), and eckols (with a dibenzodioxin linkage) (Robic et al. 2009a).

2.6.3 CAROTENOIDS

Carotenoids are isoprenoid polyene pigments that exhibit orange, purple, red, or yellow colors. They are present in chloroplasts and have been reported in bacteria, fungi, algae, and higher plants. The primary functions of carotenoids in photosynthetic organisms are related to light absorption. They assist in various processes such as photosynthesis, photo-protection, phototropism, and photoreception. Additionally, carotenoids can act as natural repellents, providing protective effects (Dembitsky and Maoka 2007; Matsuno 2001). The bioactive properties of carotenoids are determined by the structure of their polyene chains. The presence of a higher number of conjugated double bonds and specific cyclic end groups within the polyene chains play a crucial role in determining the carotenoids' ability to quench singlet oxygen and scavenge free radicals (Britton 1995).

Fucoxanthin is a major well-known compound belonging to the group of carotenoids. Fucoxanthin is the most abundant of all carotenoids, accounting for more than 10% of the estimated total natural production of carotenoids (Matsuno 2001). The global fucoxanthin market size was valued at US$112.37 million in 2022 and is expected to expand at a CAGR of 3.74% during the forecast period, reaching US$140.09 million by 2028. These numbers reflect the industrial importance of fucoxanthin, especially as a food or cosmetic ingredient. Fucoxanthin, known for its established bioactive properties, is currently utilized as a functional ingredient in different industries, including functional food, cosmeceuticals, and nutraceuticals. Seaweeds such as *Laminaria japonica*, *Undaria pinnatifida*, and *Sargassum fusiforme* are found to produce considerable amounts of fucoxanthin in their chloroplasts (Lourenço-Lopes et al. 2021). The unique properties of fucoxanthin make it a popular ingredient in the aforementioned industries. Fucoxanthin offers potential health benefits and has gained attention for its antioxidant, anti-inflammatory, anti-obesity, and anticancer properties. The functional food industry incorporates fucoxanthin into products to enhance their nutritional profile and promote overall well-being. In the cosmeceutical industry, fucoxanthin is used as an active ingredient in skincare and beauty products due to its potential effects on skin health and appearance. Furthermore, the nutraceutical industry employs fucoxanthin for its perceived health-promoting properties, often in the form of dietary supplements. Overall, the bioactive properties of fucoxanthin have led to its widespread application in functional food, cosmeceutical, and nutraceutical industries (Peng et al. 2011). Figure 2.9 shows the general chemical structures of fucoxanthin isolated from seaweeds.

2.6.4 STEROLS

Sterols are one of the important compounds in eukaryotic cell membranes. Moreover, cholesterols are major sterols of vertebrate organisms and a mixture of phytosterols contained in photosynthetic organisms. The conversion of farnesyl diphosphate into squalene activates the isoprenoid pathway, which stimulates sterol production in biological systems (Hartmann 1998; Lopes et al. 2013). Seaweeds are an important source of bioactive sterol. It has been reported that sterols biosynthesis in green (ergosterol), red (cholesterol and its derivatives), and brown (fucosterols and its derivatives) have the potential to develop as a functional material in different industries such as functional food or nutraceuticals (Kim and Van Ta 2011).

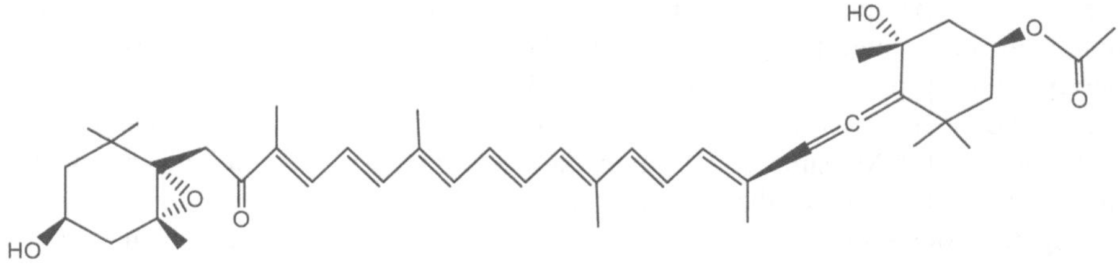

FIGURE 2.9   Chemical structure of fucoxanthin.

2.7 LIMITATIONS RELATED TO THE BIOACTIVE COMPOUNDS ISOLATED FROM SEAWEEDS

The properties of bioactive secondary metabolites found in brown seaweeds have great potential for various applications. However, scientists need to address the following challenges to take full advantage of them.

2.7.1 PRESENCE OF HEAVY METALS

One of the main challenges is the presence of heavy metals, such as Cr, Cu, Pb, Zn, Ni, Hg, As, and Cd, reported from edible seaweeds (Boucetta et al. 2019). Compared to lower-level organisms within the aquatic food chain, seaweeds are known to accumulate heavy metals at levels that can be 'thousands of times higher.' As indicated by previous studies, metal concentrations in various seaweed species were found to exceed the concentrations present in the surrounding seawater (Lindenmayer et al. 2023). Thus, edible seaweed with high levels of heavy metals can be toxic to humans, making high-level purification techniques necessary.

2.7.2 VALIDATION OF BIOACTIVE PROPERTIES

Despite the increasing interest in using bioactive compounds found in seaweed for value-added industries, validating comprehensive *in vivo* processes has been difficult due to a lack of bioavailability research. *In vitro* investigations can also make it difficult to identify the genuine potential of a compound in formulation development. Therefore, there is a need for studies to clarify the mechanism of phlorotannins *in vivo* and clinically.

2.7.3 THE TOXIC NATURE OF SOLVENTS AND HIGH COST

Ethanol, methanol, ethyl acetate, and acetone are commonly used solvents for the extraction of bioactive secondary metabolites such as phlorotannins and sterols, but these solvents are not safe for animals or humans. Researchers must be aware of the maximum residual limit of solvents. High-temperature extraction of seaweed-based bioactive secondary metabolites or microwave extraction can increase extraction yield but may also reduce the bioactive properties of extracted samples. Proper extraction conditions must be investigated. Bioactive secondary metabolites are safe and ecologically friendly approaches. However, the rate of enzymes, limited availability of substrate-specific enzymes, and limitations in maintaining bioreactor conditions require an appropriate extraction methods.

2.7.4 STRUCTURAL CHARACTERIZATION OF PURIFIED COMPOUNDS

There is lack of libraries available for comparison with standards. Spectrophotometer-based methods, such as the FRAP assay, Folin–Ciocâlteu assay, and ORAC assay, have low specificity as non-phenolic substances can overestimate results. GC-MS technology can detect pure compounds, but its limits arise when the analytical technique is more sensitive than the assay used to evaluate biological activity.

2.7.5 IMPROVING THE YIELD AND CHARACTERISTICS
OF THE SEAWEED-BASED PRODUCT

The successful marketing of a consumer product is closely tied to its cost-effective large-scale production. Many promising substances identified from seaweeds that show potential in laboratory testing fail to reach the market due to their high production costs compared to alternatives derived from other organisms, or synthetic sources (Wang et al. 2020). Therefore, further studies should be conducted to assess the economic feasibility of pilot-scale seaweed conversion facilities in the food industry. Such research is essential for providing a clear evaluation of the potential industrial applicability of seaweed-based products, thereby contributing to the development of sustainable and economically viable solutions for the food industry.

2.8   FUTURE DIRECTIONS

To fully harness the potential of seaweeds as food materials or functional products, it is crucial to develop processes that can provide biomass compatible with market demand, ensuring high yields and increased competitiveness. This objective will lead to the advancement of genetic and metabolic engineering techniques, with a particular focus on seaweeds. Optimizing the culture conditions of seaweeds not only affects the quantity of biomass produced but also influences the type and quantity of secondary metabolites. Furthermore, advancements in extraction processes and bioreactor structures are required to scale up production volumes by mimicking the marine environment to maximize productivity.

REFERENCES

Agrawal, S. C. 2012. "Factors controlling induction of reproduction in algae – review: The text." *Folia Microbiologia (Praha)* 57 (5):387–407. doi: 10.1007/s12223-012-0147-0

Anderson, L. W. J. 2007. "Control of invasive seaweeds." *Botanica Marina* 50 (5–6):418–437. doi: 10.1515/bot.2007.045

Andrade, L. R., L. T. Salgado, M. Farina, M. S. Pereira, P. A. Mourao, and G. M. Amado Filho. 2004. "Ultrastructure of acidic polysaccharides from the cell walls of brown algae." *Journal of Structural Biology* 145 (3):216–225. doi: 10.1016/j.jsb.2003.11.011

Asanka, S. K. K., H.-S. Kim, and Y.-J. Jeon. 2018. "Edible Korean Seaweed: A Source of Functional Compounds." In *Korean Functional Foods*, edited by K.-Y. Park, D. Y. Kwon, K. W. Lee, and S. Park, 359–384. CRC Press.

Baweja, P., S. Kumar, D. Sahoo, and I. Levine. 2016. "Biology of Seaweeds." In *Seaweed in Health and Disease Prevention*, edited by J. Fleurence and I. Levine, 41–106. San Diego: Academic Press.

Berteau, O., and B. Mulloy. 2003. "Sulfated fucans, fresh perspectives: Structures, functions, and biological properties of sulfated fucans and an overview of enzymes active toward this class of polysaccharide." *Glycobiology* 13 (6):29R–40R. doi: 10.1093/glycob/cwg058

Blondin, C., and A. de Agostiniz. 1995. "Biological activities of polysaccharides from marine algae." *Drugs of the Future* 20 (12):1237–1249.

Boucetta, S., W. Benchalel, S. Ferroudj, Z. Bouslama, and H. Elmsellem. 2019. "Trace metal biomonitoring of algae (*Ulva lactuca*), and mollusks (*Patella caerulea*; *Stramonita haemastoma*; *Phorcus turbinatus*) along the Eastarn-Algerian coast."

Moroccan Journal of Chemistry 7 (3):444–459. doi: 10.48317/IMIST.PRSM/morjchem-v7i3.17225

Britton, G. 1995. "Structure and properties of carotenoids in relation to function." *The FASEB Journal* 9 (15):1551–1558.

Cai, J., A. Lovatelli, J. Aguilar-Manjarrez, L. Cornish, L. Dabbadie, A. Desrochers, S. Diffey, E. G. Gamarro, J. Geehan, and A. Hurtado. 2021. Seaweeds and microalgae: An overview for unlocking their potential in global aquaculture development. FAO Fisheries and Aquaculture Circular No. 1229. Rome, Italy: FAO.

Callow, J. A. 1985. "Sexual recognition and fertilization in brown algae." *Journal of Cell Science* 2 (Supplement 2):219–232. doi: 10.1242/jcs.1985.supplement_2.12

Chapman, V. 2012. *Seaweeds and Their Uses*. The Netherlands: Springer Science + Business Media.

Charrier, B., M. H. Abreu, R. Araujo, A. Bruhn, J. C. Coates, O. De Clerck, C. Katsaros, R. R. Robaina, and T. Wichard. 2017. "Furthering knowledge of seaweed growth and development to facilitate sustainable aquaculture." *New Phytologist* 216 (4):967–975. doi: 10.1111/nph.14728

Chen, Y., Y. Yin, and J. Wang. 2022. "Biohydrogen production using macroalgal biomass of *Laminaria japonica* pretreated by gamma irradiation as substrate." *Fuel* 309. doi: 10.1016/j.fuel.2021.122179

Cheng, H., X. Zhang, Z. Cui, and S. Mao. 2021. "Grafted Polysaccharides as Advanced Pharmaceutical Excipients." In *Advances and Challenges in Pharmaceutical Technology*, edited by A. K. Nayak, K. Pal, I. Banerjee, S. Maji, and U. Nanda, 75–129. Academic Press.

Clayton, M. N. 1992. "Propagules of marine macroalgae: Structure and development." *British Phycological Journal* 27 (3):219–232. doi: 10.1080/00071619200650231

Corino, C., S. C. Modina, A. Di Giancamillo, S. Chiapparini, and R. Rossi. 2019. "Seaweeds in pig nutrition." *Animals* 9 (12):1126.

de Bettignies, T., T. Wernberg, C. Frederico, and D. Gurgel. 2018. "Exploring the influence of temperature on aspects of the reproductive phenology of temperate seaweeds." *Frontiers in Marine Science* 5:218. doi: 10.3389/fmars.2018.00218

Dembitsky, V. M., and T. Maoka. 2007. "Allenic and cumulenic lipids." *Progress in Lipid Research* 46 (6):328–375. http://dx.doi.org/10.1016/j.plipres.2007.07.001

Fernando, I. P. S., T. U. Jayawardena, K. K. A. Sanjeewa, L. Wang, Y. J. Jeon, and W. W. Lee. 2018. "Anti-inflammatory potential of alginic acid from *Sargassum horneri* against urban aerosol-induced inflammatory responses in keratinocytes and macrophages." *Ecotoxicology and Environmental Safety* 160:24–31. doi: 10.1016/j.ecoenv.2018.05.024

Froehlich, H. E., J. C. Afflerbach, M. Frazier, and B. S. Halpern. 2019. "Blue growth potential to mitigate climate change through seaweed offsetting." *Current Biology* 29 (18):3087–3093.e3. doi: 10.1016/j.cub.2019.07.041

Glasson, C. R. K., C. A. Luiten, S. M. Carnachan, A. M. Daines, J. T. Kidgell, S. F. R. Hinkley, C. Praeger, M. Andrade Martinez, L. Sargison, M. Magnusson, R. de Nys, and I. M. Sims. 2022. "Structural characterization of ulvans extracted from blade (*Ulva ohnoi*) and filamentous (*Ulva tepida* and *Ulva prolifera*) species of cultivated *Ulva*." *International Journal of Biological Macromolecules* 194:571–579. doi: 10.1016/j.ijbiomac.2021.11.100

Gomez-Gutierrez, C. M., G. Guerra-Rivas, I. E. Soria-Mercado, and N. E. Ayala-Sánchez. 2011. "Marine edible algae as disease preventers." *Advances in Food and Nutrition Research*, 64 :29–39.

Groth, I., N. Grünewald, and S. Alban. 2009. "Pharmacological profiles of animal- and nonanimal-derived sulfated polysaccharides: Comparison of unfractionated heparin, the semisynthetic glucan sulfate PS3, and the sulfated polysaccharide fraction isolated from *Delesseria sanguinea*." *Glycobiology* 19 (4):408–417.

Gupta, S., and N. Abu-Ghannam. 2011. "Bioactive potential and possible health effects of edible brown seaweeds." *Trends in Food Science & Technology* 22 (6):315–326. doi: 10.1016/j.tifs.2011.03.011

Handique, J. G. and J. B. Baruah. 2002. "Polyphenolic compounds: an overview." *Reactive and Functional Polymers* 52 (3): 163–188. doi: 10.1016/S1381-5148(02)00091-3

Harley, C. D., K. M. Anderson, K. W. Demes, J. P. Jorve, R. L. Kordas, T. A. Coyle, and M. H. Graham. 2012. "Effects of climate change on global seaweed communities." *Journal of Phycology* 48 (5):1064–1078. doi: 10.1111/j.1529-8817.2012.01224.x

Hartmann, M.-A. 1998. "Plant sterols and the membrane environment." *Trends in Plant Science* 3 (5):170–175. http://dx.doi.org/10.1016/S1360-1385(98)01233-3

Hiraoka, M., S. Shimada, M. Ohno, and Y. Serisawa. 2003. "Asexual life history by quadri-flagellate swarmers of *Ulva spinulosa* (Ulvales, Ulvophyceae)." *Phycological Research* 51 (1):29–34.

Hoffmann, A. J. 1987. "The arrival of seaweed propagules at the shore: A review." *Botanica Marina* 30 (2):151–166. doi: 10.1515/botm.1987.30.2.151

Holdt, S. L., and S. Kraan. 2011. "Bioactive compounds in seaweed: Functional food applications and legislation." *Journal of Applied Phycology* 23 (3):543–597. doi: 10.1007/s10811-010-9632-5

Jeon, Y.-J., and K. K. A. Sanjeewa. 2018b. "Edible brown seaweeds: A review." *Journal of Food Bioactives* 2:37–50. doi: 10.31665/jfb.2018.2139

Karuppusamy, S., G. Rajauria, S. Fitzpatrick, H. Lyons, H. McMahon, J. Curtin, B. K. Tiwari, and C. O'Donnell. 2022. "Biological properties and health-promoting functions of laminarin: A comprehensive review of preclinical and clinical studies." *Marine Drugs* 20 (12):772. doi: 10.3390/md20120772

Khan, F., G. J. Jeong, M. S. A. Khan, N. Tabassum, and Y. M. Kim. 2022. "Seaweed-derived phlorotannins: A review of multiple biological roles and action mechanisms." *Marine Drugs* 20 (6):384. doi: 10.3390/md20060384

Kim, S.-K., and Q. Van Ta. 2011. "Potential beneficial effects of marine algal sterols on human health." *Advances in Food and Nutrition Research* 64:191–198.

Leal, P. P., C. L. Hurd, and M. Y. Roleda. 2014. "Meiospores produced in sori of nonsporophyllous laminae of *Macrocystis pyrifera* (Laminariales, Phaeophyceae) may enhance reproductive output." *Journal of Phycology* 50 (2):400–405. doi: 10.1111/jpy.12159

Li, J., Z. He, Y. Liang, T. Peng, and Z. Hu. 2022. "Insights into algal polysaccharides: A review of their structure, depolymerases, and metabolic pathways." *Journal of Agricultural and Food Chemistry* 70 (6):1749–1765. doi: 10.1021/acs.jafc.1c05365

Lindenmayer, R., L. Lu, F. Eivazi, and Z. Afrasiabi. 2023. "Atomic spectroscopy-based analysis of heavy metals in seaweed species." *Applied Sciences* 13 (8):4764. doi: 10.3390/app13084764

Lipinska, A. P., J. Collen, S. A. Krueger-Hadfield, T. Mora, and E. Ficko-Blean. 2020. "To gel or not to gel: Differential expression of carrageenan-related genes between the gametophyte and tetrasporophyte life cycle stages of the red alga *Chondrus crispus*." *Scientific Reports* 10 (1):11498. doi: 10.1038/s41598-020-67728-6

Lopes, G., C. Sousa, P. Valentão, and P. B. Andrade. 2013. "Sterols in Algae and Health." In *Bioactive Compounds from Marine Foods*, edited by B. Hernández-Ledesma and M. Herrero, 173–191. John Wiley & Sons Ltd.

Lourenço-Lopes, C., M. Fraga-Corral, C. Jimenez-Lopez, M. Carpena, A. G. Pereira, P. Garcia-Oliveira, M. A. Prieto, and J. Simal-Gandara. 2021. "Biological action mechanisms of fucoxanthin extracted from algae for application in food and cosmetic industries." *Trends in Food Science & Technology* 117:163–181. doi: 10.1016/j.tifs.2021.03.012

Mahadevan, K. 2015. "Seaweeds: A Sustainable Food Source." In *Seaweed Sustainability*, edited by B. K. Tiwari and D. J. Troy, 347–364. San Diego: Academic Press.

Makkar, H. P. S., G. Tran, V. Heuzé, S. Giger-Reverdin, M. Lessire, F. Lebas, and P. Ankers. 2016. "Seaweeds for livestock diets: A review." *Animal Feed Science and Technology* 212:1–17. doi: 10.1016/j.anifeedsci.2015.09.018

Matsuno, T. 2001. "Aquatic animal carotenoids." *Fisheries Science* 67 (5):771–783. doi: 10.1046/j.1444-2906.2001.00323.x

Mayanglambam, A., and D. Sahoo. 2015. "Red Algae." In *The Algae World*, edited by D. Sahoo and J. Seckbach, 205–234. Dordrecht, The Netherlands: Springer.

Mohamed, S., S. N. Hashim, and H. A. Rahman. 2012. "Seaweeds: A sustainable functional food for complementary and alternative therapy." *Trends in Food Science & Technology* 23 (2):83–96. http://dx.doi.org/10.1016/j.tifs.2011.09.001

Moreira, A., S. Cruz, R. Marques, and P. Cartaxana. 2021. "The underexplored potential of green macroalgae in aquaculture." *Reviews in Aquaculture* 14 (1):5–26. doi: 10.1111/raq.12580

Nautiyal, O. H. 2013. "Natural products from plant, microbial and marine species." *International Journal of Science & Technology* 10 (1):611–646.

Navarro, N., P. Huovinen, and I. Gómez. 2020. "Life History Strategies, Photosynthesis, and Stress Tolerance in Propagules of Antarctic Seaweeds." In *Antarctic Seaweeds*, edited by I. Gómez and P. Huovinen, 193–215. Cham: Springer International Publishing.

Otaíza, R. D., C. Y. Rodríguez, J. H. Cáceres, and Á. G. Sanhueza. 2018. "Fragmentation of thalli and secondary attachment of fragments of the agarophyte *Gelidium lingulatum* (Rhodophyta, Gelidiales)." *Journal of Applied Phycology* 30 (3):1921–1931. doi: 10.1007/s10811-018-1391-8

Park, E., H. Yu, J. H. Lim, J. Hee Choi, K. J. Park, and J. Lee. 2023. "Seaweed metabolomics: A review on its nutrients, bioactive compounds and changes in climate change." *Food Research International* 163:112221. doi: 10.1016/j.foodres.2022.112221

Peng, Y., J. Hu, B. Yang, X.-P. Lin, X.-F. Zhou, X.-W. Yang, and Y. Liu. 2015. "Chemical Composition of Seaweeds." In *Seaweed Sustainability*, edited by B. K. Tiwari and D. J. Troy, 79–124. San Diego: Academic Press.

Peng, J., J.-P. Yuan, C.-F. Wu, and J.-H. Wang. 2011. "Fucoxanthin, a marine carotenoid present in brown seaweeds and diatoms: Metabolism and bioactivities relevant to human health." *Marine Drugs* 9 (10):1806.

Pereira, L. 2021. "Macroalgae." *Encyclopedia* 1 (1):177–188. doi: 10.3390/encyclopedia1010017

Rashedy, S. H., M. S. M. Abd El Hafez, M. A. Dar, J. Cotas, and L. Pereira. 2021. "Evaluation and characterization of alginate extracted from brown seaweed collected in the Red Sea." *Applied Sciences* 11 (14):6290.

Relleve, L., N. Nagasawa, L. Q. Luan, T. Yagi, C. Aranilla, L. Abad, T. Kume, F. Yoshii, and A. dela Rosa. 2005. "Degradation of carrageenan by radiation." *Polymer Degradation and Stability* 87 (3):403–410. doi: 10.1016/j.polymdegradstab.2004.09.003

Rioux, L. E., S. L. Turgeon, and M. Beaulieu. 2007a. "Characterization of polysaccharides extracted from brown seaweeds." *Carbohydrate Polymers* 69 (3):530–537. doi: 10.1016/j.carbpol.2007.01.009

Rioux, L.-E., S. L. Turgeon, and M. Beaulieu. 2007b. "Rheological characterisation of polysaccharides extracted from brown seaweeds." *Journal of the Science of Food and Agriculture* 87 (9):1630–1638. doi: 10.1002/jsfa.2829

Robic, A., D. Bertrand, J. F. Sassi, Y. Lerat, and M. Lahaye. 2009a. "Determination of the chemical composition of ulvan, a cell wall polysaccharide from *Ulva* spp. (Ulvales, Chlorophyta) by FT-IR and chemometrics." *Journal of Applied Phycology* 21 (4): 451–456. doi: 10.1007/s10811-008-9390-9

Robic, A., C. Rondeau-Mouro, J. F. Sassi, Y. Lerat, and M. Lahaye. 2009b. "Structure and interactions of ulvan in the cell wall of the marine green algae *Ulva rotundata* (Ulvales, Chlorophyceae)." *Carbohydrate Polymers* 77 (2):206–216. http://dx.doi.org/10.1016/j.carbpol.2008.12.023

Robic, A., J. F. Sassi, and M. Lahaye. 2008. "Impact of stabilization treatments of the green seaweed *Ulva rotundata* (Chlorophyta) on the extraction yield, the physico-chemical and rheological properties of ulvan." *Carbohydrate Polymers* 74 (3):344–352. http://dx.doi.org/10.1016/j.carbpol.2008.02.020

Sanjeewa, K. K. A., K. Herath, H. W. Yang, C. S. Choi, and Y. J. Jeon. 2021. "Anti-inflammatory mechanisms of fucoidans to treat inflammatory diseases: A review." *Marine Drugs* 19 (12):678. doi: 10.3390/md19120678

Sanjeewa, K. K. A., T. U. Jayawardena, S.-Y. Kim, H.-S. Kim, G. Ahn, J. Kim, and Y.-J. Jeon. 2019. "Fucoidan isolated from invasive *Sargassum horneri* inhibit LPS-induced inflammation via blocking NF-κB and MAPK pathways." *Algal Research* 41:101561. doi: 10.1016/j.algal.2019.101561

Sanjeewa, K. K. A., J. S. Lee, W. S. Kim, and Y. J. Jeon. 2017. "The potential of brown-algae polysaccharides for the development of anticancer agents: An update on anticancer effects reported for fucoidan and laminaran." *Carbohydrate Polymers* 177:451–459. doi: 10.1016/j.carbpol.2017.09.005

Senni, K., J. Pereira, F. Gueniche, C. Delbarre-Ladrat, C. Sinquin, J. Ratiskol, G. Godeau, A.-M. Fischer, D. Helley, and S. Colliec-Jouault. 2011. "Marine polysaccharides: A source of bioactive molecules for cell therapy and tissue engineering." *Marine Drugs* 9 (9):1664–1681.

Shim, E., H. Park, S. H. Im, G. C. Zuccarello, and G. H. Kim. 2020. "Dynamics of spermatial nuclei in trichogyne of the red alga *Bostrychia moritziana* (Florideophyceae)." *Algae* 35 (4):389–404. doi: 10.4490/algae.2020.35.12.7

Smith, A. J., B. Graves, R. Child, P. J. Rice, Z. Ma, D. W. Lowman, H. E. Ensley, K. T. Ryter, J. T. Evans, and D. L. Williams. 2018. "Immunoregulatory activity of the natural product laminarin varies widely as a result of its physical properties." *Journal of Immunology* 200 (2):788–799. doi: 10.4049/jimmunol.1701258

Sudhakar, K., R. Mamat, M. Samykano, W. H. Azmi, W. F. W. Ishak, and Talal Yusaf. 2018. "An overview of marine macroalgae as bioresource." *Renewable and Sustainable Energy Reviews* 91:165–179. doi: 10.1016/j.rser.2018.03.100

Sultana, F., M. A. Wahab, M. Nahiduzzaman, M. Mohiuddin, M. Z. Iqbal, A. Shakil, A.-A. Mamun, M. S. R. Khan, L. Wong, and M. Asaduzzaman. 2023. "Seaweed farming for food and nutritional security, climate change mitigation and adaptation, and women empowerment: A review." *Aquaculture and Fisheries* 8 (5):463–480. doi: 10.1016/j.aaf.2022.09.001

Syakilla, N., R. George, F. Y. Chye, W. Pindi, S. Mantihal, N. A. Wahab, F. M. Fadzwi, P. H. Gu, and P. Matanjun. 2022. "A review on nutrients, phytochemicals, and health benefits of green seaweed, *Caulerpa lentillifera*." *Foods* 11 (18):2832. doi: 10.3390/foods11182832

Synytsya, A., W.-J. Kim, S.-M. Kim, R. Pohl, A. Synytsya, F. Kvasnička, J. Čopíková, and Y. I. Park. 2010. "Structure and antitumour activity of fucoidan isolated from sporophyll of Korean brown seaweed *Undaria pinnatifida*." *Carbohydrate Polymers* 81 (1):41–48. doi: http://dx.doi.org/10.1016/j.carbpol.2010.01.052

Wang, S., S. Zhao, B. B. Uzoejinwa, A. Zheng, Q. Wang, J. Huang, and A. E.-F. Abomohra. 2020. "A state-of-the-art review on dual purpose seaweeds utilization for wastewater treatment and crude bio-oil production." *Energy Conversion and Management* 222:113253. doi: 10.1016/j.enconman.2020.113253

Xiaojun, Y., L. Xiancui, Z. Chengxu, and F. Xiao. 1996. "Prevention of fish oil rancidity by phlorotannins from *Sargassum kjellmanianum*." *Journal of Applied Phycology* 8 (3):201–203. doi: 10.1007/BF02184972

Zechman, F. W., and A. C. Mathieson. 1985. "The distribution of seaweed propagules in estuarine, coastal and offshore waters of New Hampshire, U.S.A." *Botanica Marina* 28 (7):283–294. doi: 10.1515/botm.1985.28.7.283

Zia, K. M., S. Tabasum, M. Nasif, N. Sultan, A. Aslam, A. Noreen, and M. Zuber. 2017. "A review on synthesis, properties and applications of natural polymer based carrageenan blends and composites." *International Journal of Biological Macromolecules* 96: 282–301. doi: 10.1016/j.ijbiomac.2016.11.095

3 Marine Bacteria and Cyanobacteria

3.1 INTRODUCTION TO MARINE MICROORGANISMS

Microorganisms play a vital role in the preservation and sustainability of ecosystems as they possess the ability to adapt quickly to environmental changes and degradation. In general, microorganisms grow in a wide range of environments, including harsh ecosystems such as volcanic eruptions and Antarctic glaciers (Dash et al. 2013). Hence, it is evident that microorganisms are omnipresent, including within marine environments. These microorganisms showcase their resilience and capacity to thrive in diverse marine habitats, contributing significantly to the intricate balance and functioning of marine ecosystems.

Other than the unique role of marine microorganisms in these marine ecosystems, marine microorganisms have also gained increasing attention in biodiscovery due to their unique physicochemical characteristics developed in response to extreme conditions in the marine environment. Studies have highlighted the potential of these microorganisms and their metabolites as lead compounds in pharmaceutical development, biofuel, and functional items, including cosmeceuticals and functional foods (Ameen, AlNadhari and Al-Homaidan 2021; Wagner-Döbler et al. 2002). This chapter aims to explore the potential of marine bacteria and cyanobacteria and their prospective applications in modern nutraceuticals and functional foods, considering their sustainable nature and potential as food and pharmaceutical ingredients.

3.2 MAJOR TYPES OF MARINE MICROORGANISMS

The oceans are home to various habitats that exhibit diverse conditions: different temperatures, pH, currents, different salinity levels, wind patterns, and precipitation regimes. With these varying environmental conditions, the microorganisms' habitats in marine ecosystems are adapted to live in harsh marine conditions; thus, these microorganisms possess complex characteristic features of adaptation (Dash et al. 2013; Ghosh et al. 2022). Therefore, marine microorganisms are believed to possess a significant potential for producing bioactive compounds that are not typically found in terrestrial environments (Ameen, AlNadhari and Al-Homaidan 2021). Furthermore, habitats of microorganisms (both prokaryotes and eukaryotes) were reported from all marine ecosystems with an estimated abundance of 10^4–10^7 cells/ml in seawater and, 10^3–10^{10} cells/cm^3 in marine sediments. In these habitats, these microorganisms are believed to play key roles in marine biogeochemical cycles (Wang et al. 2021).

3.2.1   MARINE BACTERIA

Bacteria, being among the earliest life-forms on Earth, continue to thrive in nearly every corner of Earth, including terrestrial and marine environments (DeLong, Pace and Kane 2001). They encompass a diverse group of prokaryotic microorganisms. Bacteria are typically small, single-celled organisms, measuring only a few micrometers in length. In general, bacteria have different shapes and forms: spheres, rods, and spirals in nature (Hamidi et al. 2019). Various adaptation of marine bacteria to challenging environments has resulted in a vast biological and genetic diversity in marine bacteria. Specifically, studies have reported that over 98% of the biomass in the ocean is accounted for by marine bacteria (Zhang and Kim 2012). This diversity has sparked considerable interest in marine bacteria as valuable biotechnological resources. These bacteria hold immense potential as a source of novel bioactive compounds with applications in various industries: industrial manufacturing, agriculture, environmental remediation, pharmaceutical development, and medical research (Debnath, Paul and Bisen 2007). The exploration of marine bacteria and their bioactive compounds opens up exciting opportunities for innovative solutions in numerous fields. Some potential industries/applications that can benefit from marine bacteria are summarized in the following sections.

3.2.1.1   Industrially Useful Enzymes

In general, enzymes produced by various microorganism's habitats in wide temperature range, high pressure, extensive pH, and high salinity levels in marine ecosystems make them ideal use in environment-friendly industrial applications with minimum operational cost due to the different properties that they achieve due to the harsh environment conditions in their host's habitats. Industrial applications of marine-derived enzymes include pharmacy, food and beverages, dairy products, detergents, and the waste treatment industry (Figure 3.1).

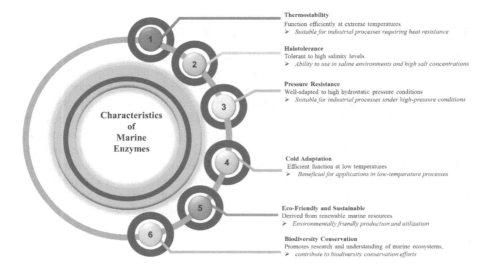

FIGURE 3.1   Major properties and advantages of marine enzymes.

Taken together, during the last few decades, enzymes derived from marine microbes have gained significant attention, particularly for their industrial applications to minimize operational costs and to achieve sustainable development goals (Barzkar and Sohail 2020). Numerous studies have demonstrated the ability of marine bacteria to produce a diverse range of industrially important enzymes. As microbial enzymes exhibit greater stability and activity compared to enzymes derived from animals or plants, there has been an increasing interest in exploring microbial sources for new enzyme discoveries. The enzymes produced by marine bacteria include agarases (Ma et al. 2007), carrageenases (Zhu and Ning 2016), xylanases (Guo et al. 2018), α-amylase (Chakraborty et al. 2014), chitinases (Gaber et al. 2016), proteases (Farhadian, Asoodeh and Lagzian 2015), lipases (Sathishkumar et al. 2015), and collagenases (Bhattacharya et al. 2019). Other than the aforementioned enzymes, cellulose-degrading enzymes identified from marine bacteria have also emerged as leading biocatalysts with great potential for use in biorefineries (Barzkar and Sohail 2020). Cellulose-degrading enzymes play a crucial role in breaking down cellulose, a complex polysaccharide found in plant biomass, into simpler sugars that can be further processed into biofuels, biochemicals, and other valuable products in the biorefinery industry.

The marine enzyme market is segmented according to its applications, which include animal feed, pharmaceuticals, leather, cosmetics, food and beverage industries, and paper. Among these industries, the food and beverage sector stands out as the most profitable in the commercial aspect. However, it is expected that enzymes identified from marine microorganisms will soon find broader applications across different industries due to their favorable characteristics and growing popularity in the food industry. Furthermore, the marine enzymes market for the pharmaceutical industry is also expected to grow within the next few years (Ghattavi and Homaei 2023).

To enhance current practices, several key changes are required. These include the complete purification of enzymes derived from marine organisms, identification of their active sites, and advancements in enzyme immobilization techniques. Improvements and developments in these areas will contribute to harnessing the full potential of marine enzymes for biological processes, biological and biochemical sensors, and other applications (Figure 3.2) (Ghattavi and Homaei 2023). In the context of future biotechnological applications, a comprehensive understanding of the relationship between the structure and performance of biocatalysts is essential for their successful commercialization.

3.2.1.2   Role of Marine Bacteria in Aquaculture

Aquaculture is one of the fastest-growing sectors in the food production industry and In 2018, global aquaculture fish production reached 82.1 million tons, which accounted for 46% of global fish production (FAO 2020).

However, intensive fish farming practices have been associated with an increased likelihood of disease occurrence, which can lead to reduced production rates and compromised flesh quality, ultimately impacting the global economy. Antibiotics have traditionally been favored by farmers due to their fast-acting nature and easy accessibility. Moreover, excessive use of antibiotics was found to demonstrate detrimental effects on the aquatic ecosystem (Banerjee and Ray 2017; Cabello 2006).

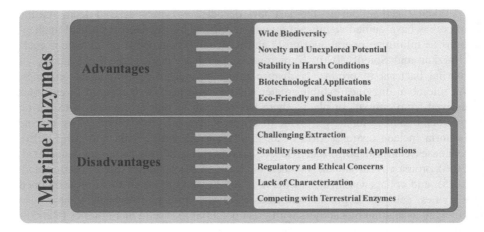

FIGURE 3.2 The advantages and disadvantages of using marine enzymes.

As an alternative approach to maintaining ecosystem health and controlling fish diseases, the substitution of antibiotics with probiotics has gained attention (Martinez Cruz et al. 2012).

As a result, the evaluation of probiotic bacteria capable of producing bacteriocins (Box 3.1) is an area of extensive research across multiple sectors, including aquaculture (Martinez Cruz et al. 2012). The rationale behind utilizing marine-based bacteriocins in aquaculture stems from the fact that the bacterial strains producing them occupy similar ecological niches as the targeted pathogens. In contrast, the use of terrestrial bacterial strains as probiotics for aquaculture has demonstrated limited success due to the dependence of bacterial strain characteristics on their specific environmental conditions. Therefore, a more effective approach is to isolate bacteriocin-producing probiotic bacteria from the marine environment, where they can thrive optimally. These marine strains are anticipated to be more efficient in aquaculture applications compared to non-marine bacteria (Rather et al. 2017). Even though the initial interest was focused on marine bacteria as growth promoters, new areas have been found to improve the health of aquacultured animals, such as their effect on reproduction or stress tolerance, although this requires a more scientific development (Huang et al. 2008; Martinez Cruz et al. 2012).

BOX 3.1 BACTERIOCINS

Bacteriocins are antibiotics produced by strains of certain species of microorganisms that are active against other strains of the same or related species. They can function as natural food preservatives through the inhibition of spoilage or pathogenic bacteria and ultimately contribute to food safety.

(Gundogan 2014)

3.2.1.3   Role of Marine Bacteria in Heavy Metal Bioremediation

There is no widely agreed-upon criterion-based definition for the term 'heavy metal'. Different meanings may be attached to the term, depending on the context (Orekhova, Davydova and Smirnov 2023). In general, heavy metals are characterized as metal ions with a density exceeding 5 g/cm^3 and most of the metal ions that come under the heavy metal category are toxic and carcinogenic to humans even at low concentrations (Shrestha et al. 2021). Furthermore, these heavy metals are nonbiodegradable, leading to their tendency to accumulate in living organisms over time, resulting in an increase in their concentration (Cabral Pinto et al. 2017). Arsenic, boron, cobalt, copper, cadmium, chromium, vanadium, mercury, molybdenum, nickel, strontium, titanium, tin, lead, zinc, etc. have been identified as major heavy metal contaminants in different ecosystems due to their extensive use in different industries such as plating and electroplating industry, pesticides, rayon industry, mining industry, metal smelting, metal rinse processes, tanning industry, fluidized bed bioreactors, batteries, textile industry, petrochemicals, paper manufacturing, and electrolysis applications (Kar et al. 2007; Qasem, Mohammed and Lawal 2021; Vardhan, Kumar and Panda 2019). Among those, Hg^{2+}, Pb^{2+}, and Cd^{2+} are further identified as major heavy metal pollutants due to their toxicity, nonbiodegradability, and availability in a variety of waste streams (De, Ramaiah and Vardanyan 2008).

Over the past few decades, there have been extensive efforts to develop effective and cost-efficient approaches for the removal of heavy metals from wastewater used for different industrial applications (Mishra et al. 2008). Conventional methods like chemical precipitation, adsorption, ion flotation, ion exchange, coagulation/flocculation, and electrochemical techniques are some known examples of conventional heavy metal removal techniques (Shrestha et al. 2021). However, these methods have exhibited certain drawbacks, including substantial sludge production necessitating additional treatment, limited removal efficiency, and high energy demands (Zamora-Ledezma et al. 2021). In recent years, newer and more efficient technologies with enhanced economic viability and innovative approaches are under investigation. Notably, photocatalysis, electrodialysis, hydrogels, membrane separation techniques, and the introduction of novel adsorbents have emerged as potential alternatives to achieve improved adsorption capabilities (Figure 3.3) (Zamora-Ledezma et al. 2021). In this context, the advancement of new technologies has given rise to the emergence of 'Bioremediation' as a potent alternative approach aimed at mitigating the detrimental effects of substantial heavy metal accumulation.

Bioremediation is a technology that harnesses the metabolic capabilities of microorganisms to remediate polluted environments (Mishra, Singh and Kumar 2021). Biodegradation processes can convert organic contaminants, leading to either complete transformation into inorganic products (mineralization) or changes in the transport of inorganic contaminants. In certain instances, natural intrinsic bioremediation may be adequate for reducing risks (Seagren 2023). One key aspect of bioremediation is its implementation in natural, nonsterile environments that host diverse organisms. Bacteria, particularly those with pollutant-degrading abilities, play essential roles in bioremediation, but other organisms such as fungi and grazing protozoa also influence the process (Watanabe 2001). Compared to terrestrial ecosystems, marine environments are more susceptible to various

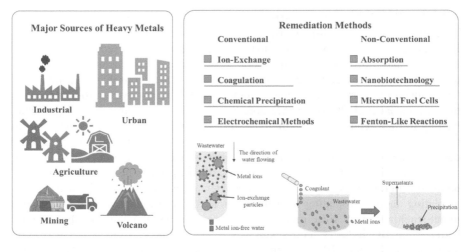

FIGURE 3.3   Heavy metals–related health hazards and heavy metal treatment methods.

pollutants, including oil spills, heavy metals, organic contaminants, plastic waste, and microplastic, which can have negative effects on marine ecosystems (Sharma and Chatterjee 2017). Implementing bioremediation techniques is an environment-friendly and cost-effective alternative to traditional physical removal methods (Sivaperumal, Kamala and Rajaram 2017).

When considering about potential microorganisms for bioremediation purposes, marine microbes possess a higher potential for use in bioremediation applications. Specifically, adaptability to adverse conditions is due to the constant environmental variations (pH, temperature, currents precipitation regimes, salinity, sea surface temperature, and wind pattern) in their habitat. This adaptability is a significant advantage when considering their *in situ* use for bioremediation purposes. In addition, utilizing bacteria from deep marine environments in bioremediation processes offers a direct application of these microbes in challenging conditions as those organisms are adapted to live in extreme environmental conditions in deep ecosystems (Singh, Goutam and Ghosh 2022). Deep-marine bacteria employ various strategies to withstand high concentrations of metals, including extracellular sequestration, precipitation, reduction, efflux mechanisms, biosorption, altered permeability, and intracellular bioaccumulation. These unique characteristics make deep-marine bacteria ideal candidates for the bioremediation of metal-contaminated environmental sites (Kubicki et al. 2019; Singh, Goutam and Ghosh 2022) (Table 3.1).

A number of recent studies highlighted the potential of marine bacteria for use as microorganisms in heavy metal bioabsorption programs (Fulke, Kotian and Giripunje 2020). According to Ameen, Hamdan and El-Naggar (2020), lactic acid bacteria separated from the Alexandrian Mediterranean Seacoast, Egypt, called *Lactobacillus plantarum* MF042018, was found to exhibit a high degree of resistance against nickel and chromium, at concentrations of 500 ppm and 100 ppm. Moreover,

TABLE 3.1

Bacteriocins Isolated from Different Marine Bacteria and Their Modes of Action

No.	Producing Strain	Place Collected	Mode of Action/Role	Reference
			Role of marine bacteria in aquaculture	
1	*Bacillus licheniformis,* strain BTHT8	Not specified	Inhibits the growth of *Serratia marcescens, Salmonella* spp., *Bacillus thuringiensis, Staphylococcus aureus, Escherichia coli, Enterobacter aerogenes*	Ahmad et al. (2013)
2	*Lactococcus lactis* TW34	The intestine of the fish *Odontesthes platensis*	Inhibits fish pathogen *Lactococcus garvieae*	Sequeiros et al. (2015)
3	*Bacillus licheniformis WIT 562, 564, and 566*	From seaweeds	Production of lichenicidin	Prieto et al. (2012)
			Role of marine bacteria in heavy metal bioremediation	
4	*Lactobacillus plantarum* MF042018	Seawater samples	Heavy metal absorption	Ameen, Hamdan and El-Naggar (2020)
5	*Halomonas* sp. SZN1, *Alcanivorax* sp. SZN2, *Pseudoalteromonas* sp. SZN3, *Epibacterium* sp. SZN4, and *Virgibacillus* sp. SZN7	Polluted sediments	Heavy metal immobilization activity	Dell'Anno et al. (2020)
6	*Pseudomonas chengduensis* PPSS-4	Marine sediment	Remove heavy metals by uptaking them into the cells	Priyadarshanee and Das (2021)
7	*Streptomyces rochei* ANH	Seacoast	Heavy metal absorption	Hamdan, Abd-El-Mageed and Ghanem (2021)
8	*Pseudomonas aeruginosa* N6P6	Not specified	Heavy metal absorption	Kumari and Das (2019)

according to the authors, after 24 hours of incubation, *L. plantarum* MF042018 (1% freshly prepared) with a concentration of 500 ppm for Ni or 100 ppm for Cr was able to efficiently remove Ni^{2+} and Cr^{2+} from the broth medium by $33.8 \pm 0.8\%$ and $30.2 \pm 0.5\%$, respectively (Ameen, Hamdan and El-Naggar 2020). In another study, Dell'Anno et al. (2020) attempted to evaluate the heavy metal immobilization capacity of five bacterial strains (*Halomonas* sp. SZN1, *Alcanivorax* sp. SZN2, *Pseudoalteromonas* sp. SZN3, *Epibacterium* sp. SZN4, and *Virgibacillus* sp. SZN7) isolated from polluted sediments from an abandoned industrial site in the Gulf of Naples, Mediterranean Sea (Dell'Anno et al. 2020). According to the results, a mixture of SZN1 and SZN2 was found to be effective in reducing the amount of metals associated with the carbonate/exchangeable fraction, with a reduction of 40%, 73%, and 53% for As, Pb, and Cd, respectively. In addition, different combinations of bacterial isolates were found effective against heavy metals like As, Cd, and Pb. In another study, Priyadarshanee and Das (2021) also reported the biofilm-forming and heavy metal–resistant properties of the marine bacterial strain *Pseudomonas chengduensis* PPSS-4 isolated from the contaminated marine sediment of Paradip Port, Odisha, India (Priyadarshanee and Das 2021). According to the authors, under the tested conditions, *P. chengduensis* PPSS-4 was found to possess promising properties for metal iron absorption in heavy metal–contaminated (Pb, Cr, and Cd) ecosystems. In addition, Kumari and Das (2019) also reported the lead (Pb) remediation potential of a biofilm-forming marine bacterium *Pseudomonas aeruginosa* N6P6. According to the authors, upregulated expression of bmtA gene of *P. aeruginosa* N6P6 is responsible for the metal iron absorbent properties of microbial isolate (Kumari and Das 2019). Hamdan, Abd-El-Mageed and Ghanem (2021) also attempted to investigate the biosorption potential of marine bacteria (*Streptomyces rochei* ANH) isolated from the Alexandrian Mediterranean Seacoast, Egypt, with their potential use in metal remediation of industrial effluents. This study suggests that marine actinomycete isolate, *S. rochei* ANH, is a powerful microorganism for use as bioremediation agent in heavy metal (Ni^{2+}, Cu^{2+}, Pb^{2+}, Cd^{2+}, and Cr^{6+}) contaminated water bodies such as industrial wastewater (Hamdan, Abd-El-Mageed and Ghanem 2021).

3.2.1.4 Application of Marine Bacteria in Petroleum and Diesel Biodegradation

Crude oil and its derivatives rank among the most widespread environmental pollutants in marine ecosystems. In recent years, bioremediation utilizing oil-degrading bacteria has emerged as a promising and environment-friendly method for cleaning up such contamination (Lee et al. 2018). The growing urgency to address the adverse impacts of human activities on estuaries and coastal marine ecosystems by petroleum-based pollution has driven the advancement of efficient bioremediation approaches. Among these strategies, marine bacteria capable of producing biosurfactants (Box 3.2) show great promise in bio-remediating oil pollution in marine environments, making them potential candidates for enhancing oil recovery

---

BOX 3.2 SURFACTANTS

Surfactants, also known as 'surface-active agents,' are chemical compounds that decrease the surface tension between liquids or the interfacial tension between liquids and solids. These amphiphilic compounds achieve this by replacing high-energy bulk molecules at interfaces, thus reducing the overall free energy of the system.

TABLE 3.2
Applications of Marine Bacteria as Biosurfactant

No.	Application	Place Collected	Name of Bacteria	Reference
1	Petroleum biodegradation	Coastal line of Chennai harbor	*Pseudomonas mendocina* ADY2b	Balakrishnan et al. (2022)
2	Petroleum biodegradation	Seawater, sediment	*Alcanivorax borkumens*	Ilhami and Holifah (2022)
3	Petroleum biodegradation	Seawater collected from Ny-Ålesund Harbor and the Kronebreen glacier	*Marinobacter antarcticus*	Crisafi et al. (2016)
4	Petroleum biodegradation	Seawater collected from Ny-Ålesund Harbor and the Kronebreen glacier	*Oleispira antarctica*	Crisafi et al. (2016)
5	Petroleum biodegradation	Seawater collected from Ny-Ålesund Harbor and the Kronebreen glacier	*Sphingopyxis flavimaris*	Crisafi et al. (2016)
6	Petroleum biodegradation	Seawater samples	*Bacillus methylotrophicus*	Chaprao et al. (2018); Pereira et al. (2019)
7	Petroleum biodegradation	Seawater samples	*Pseudomonas sihuiensis*	Pereira et al. (2019)
8	Petroleum biodegradation	Sediments of the harbor of Messina	*Oleiphilus messinensis*	Toshchakov et al. (2017)
9	Petroleum biodegradation	Seawater samples	*Marinobacter hydrocarbonoclasticus*	Mounier et al. (2014)
10	Petroleum biodegradation	Seabed mud	*Rhodococcus erythropolis*	Huang et al. (2008); Peng et al. (2007)

(Hassan et al. 2022; Kubicki et al. 2019). Some common bacteria identified with biosurfactant potentials are listed in Table 3.2.

3.2.1.5 Role of Marine Bacteria in Degradation of Plastic

Plastic, despite its unsustainable consumption pattern, has become an inseparable component of human life (Danso, Chow and Streit 2019b). With its cost-effective production, user-friendliness, durability, and high demand in the market, global plastic production has surged dramatically, exceeding 390.7 million tons in 2021 (Inc. 2022). The primary polymers produced around the globe with economic significance include polyurethane, polyethylene, polyamide, polyethylene terephthalate, polystyrene, polyvinylchloride, and polypropylene (Figure 3.4) (Danso, Chow and Streit 2019a; Geyer, Jambeck and Law 2017). As plastic production and consumption continue to rise, it is estimated that 5–13 million metric tons of plastic find their way into the ocean ecosystems each year through different ways such as freshwater streams, stormwater runoff, wastewater treatment plant discharge, and atmospheric transportation (Danso, Chow and Streit 2019a; Liu et al. 2019). This accumulated waste plastic carries negative consequences for diverse ecosystems, as well as poses risks to the well-being of

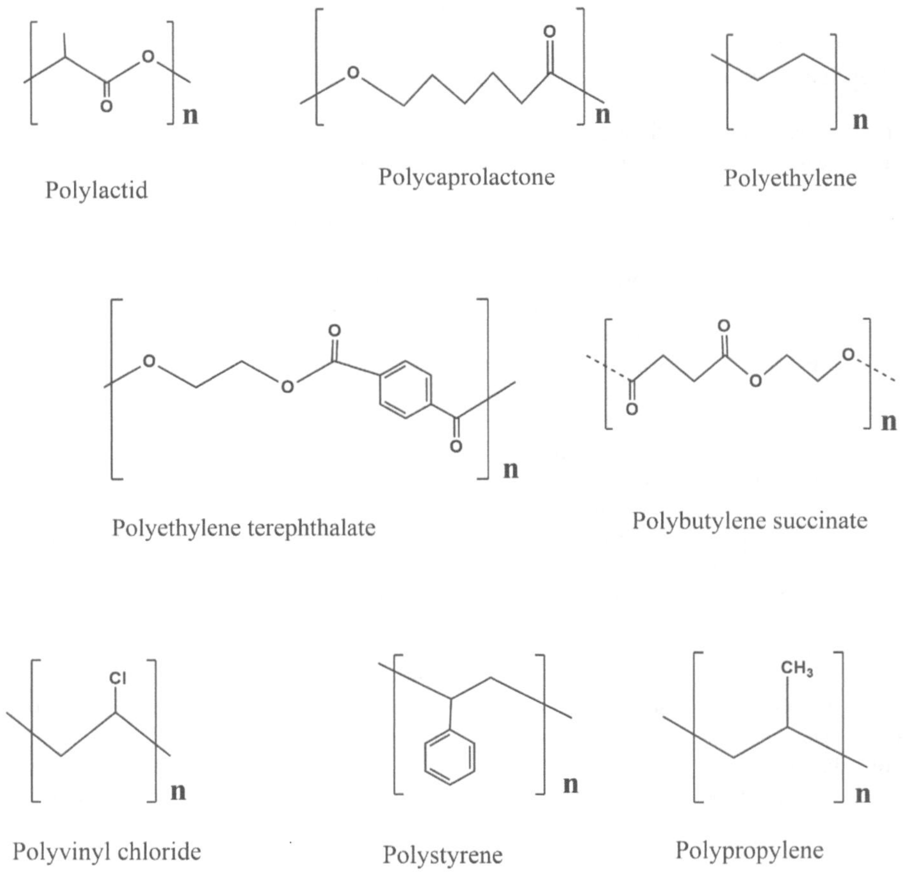

FIGURE 3.4   Structures of commonly available plastics.

both aquatic and terrestrial life-forms worldwide (Hohn et al. 2020; Joshi et al. 2022; Mora-Teddy and Matthaei 2019). Once in the ocean, larger plastic items break down into micro- and nano-plastics. Large plastic debris can directly cause the death of larger marine organisms through scenarios like entanglement, strangulation, choking, and even inducing starvation due to a misguided sense of satiation (Anastasopoulou and Fortibuoni 2019; Napper and Thompson 2019). On the other hand, smaller plastic fragments, such as micro- and nano-plastics, exert detrimental effects on marine organisms due to their significant surface-area-to-volume ratio and their capability to move within the internal organism (Tuuri and Leterme 2023).

Conventional solid waste management, such as incineration, inadvertently leads to air pollution, releasing harmful pollutants like dioxins, furans, mercury, and polychlorinated biphenyls (Verma et al. 2016; Yusuf et al. 2022). Additionally, incinerating plastic waste releases carcinogenic compounds like dioxins and furans into the environment in significant quantities (Verma et al. 2016). The ubiquity, persistence, and environmental hazards associated with plastic waste underscore the urgent need

BOX 3.3 PLASTISPHERE

Plastisphere refers to a varied microbial community that resides on small fragments of plastic found floating in the ocean. These communities exhibit notable differences from the surrounding water, indicating that plastic functions as a distinct habitat within the ocean.

for the development of efficient, environment-friendly, and sustainable microbial technologies for plastic degradation.

With the increasing concern over plastic pollution in the oceans, there has been a growing focus on the plastisphere (Box 3.3), which comprises microorganisms that interact with marine plastic debris (Roager and Sonnenschein 2019). Microbial communities that inhabit plastic have been studied in various oceanic regions, revealing significant distinctions from the communities present in the surrounding waters. Furthermore, researchers have isolated a number of bacterial strains capable of degrading plastic from diverse environments. This suggests that marine microorganisms may have adapted to plastic surfaces for colonization and potential degradation. In this section, the author attempted to summarize recent findings about plastic-degrading bacteria reported from marine ecosystems.

In a recent study, Pang et al. (2023) reported the plastic degradation properties of 55 bacterial and 184 fungal strains degrading polycaprolactone (PCL) in plastic waste samples from Dafeng coastal salt marshes, Jiangsu, China (Pang et al. 2023). According to the authors, bacterial isolates had the potential to degrade one or more types of plastics (polyethylene terephthalate, expanded polystyrene, polyethylene, polyurethane, polyamide, polypropylene, and polyvinyl chloride) under *in vitro* conditions. In another recent study, the potential of marine bacteria to degrade polystyrene (PS) in the mangrove ecosystem was explored by Liu et al. (2023). According to the authors, 7.73% and 2.66% PS was degraded by *Gordonia* and *Novosphingobium* in one-month time intervals, respectively (Liu et al. 2023). In another recent study, soil samples collected from the plastic waste dumped at coastal environmental sites in India (Chennai, Cuddalore, Kanchipuram, Nagapattinam, Pudukottai, Ramanathapuram, Thanjavur, Thiruvarur, Tuticorin, and Villupuram) were tested to identify high-density polyethylene (HDPE)-degrading bacteria (Sangeetha Devi et al. 2019). According to the authors, among the 248 bacterial isolates, *Bacillus* spp. and *Pseudomonas* spp. were identified as having promising HDPE-degrading properties. In a similar study, Kumari, Chaudhary and Jha (2019) also reported the HDPE-degrading potential of marine bacteria (*Bacillus* sp. AIIW2) isolated from coastal areas of Arambhada, Gujarat, India (Kumari, Chaudhary and Jha 2019). In this study, the authors determined the HDPE-, PVC-, and LDPE-degrading ability of *Bacillus* sp. AIIW2. According to the results, *Bacillus* sp. AIIW2 successfully degraded PVC, LDPE, and HDPE films at the rate of $0.26 \pm 0.02\%$, $0.96 \pm 0.02\%$, and $1.0 \pm 0.01\%$, respectively, within 90 days. However, compared to other bacterial strains such as *Gordonia* and *Novosphingobium* (Liu et al. 2023), the bacterial isolate (*Bacillus* sp. AIIW2) looks weaker as it shows poor performance compared to the recent studies. Additionally, in another study, the PE degradation potential of *Alcanivorax* sp. 24 was reported by Zadjelovic et al. (2022).

According to the authors, the marine bacterium isolates effectively utilized and incorporated weathered LDPE leachate; the bacterium also managed to decrease the molecular weight distribution (from 122 kg/mol to 83 kg/mol) and the overall mass of untouched LDPE films (by 0.9% after 34 days of incubation). Remarkably, *Alcanivorax* assimilated the isotopic signature of the original plastic and activated a wide range of metabolic pathways for breaking down aliphatic compounds (Zadjelovic et al. 2022). Based on the above case studies, marine bacteria and their enzymes have the potential to be used in plastic degradation applications and will be useful in removing waste plastic from the ecosystems quickly (Table 3.3). This will eventually help to create a cleaner and more productive environment for future generations.

TABLE 3.3

Applications of Marine Bacteria as Plastic-Degrading Agents

No.	Application	Place Collected	Name of Bacteria	Reference
1	Plastic degradation	Dafeng coastal salt marshes, Jiangsu, China	*Jonesia* and *Streptomyces* bacterial strains	Pang et al. (2023)
2	Plastic degradation	Epilittoral zone of Zi-Ni Town mangrove (N24°27′04.262″, E117°53′56.409″), Zhangzhou, China	*Gordonia* and *Novosphingobium* bacterial strains	Liu et al. (2023)
3	High-density polyethylene (HDPE) degradation	Marine dump sites in different locations in India	*Bacillus* spp. and *Pseudomonas* spp.	Sangeetha Devi et al. (2019)
4	HDPE degradation	Coastal areas of Arambhada, Gujarat, India	*Bacillus* spp. AIIW2	Kumari, Chaudhary and Jha (2019)
5	Low-density polyethylene (LDPE) degradation	From marine plastic debris	*Alcanivorax* spp. 24	Zadjelovic et al. (2022)
6	Polycyclic aromatic hydrocarbons (PAHs) degradation	Coastal regions of Tamil Nadu, India	*Klebsiella pneumonia*	Mohanrasu et al. (2018)
7	HDPE degradation	Coastal regions of Tamil Nadu, India	*Brevibacillus borstelensis*	Mohanrasu et al. (2018)
8	Polyethylene terephthalate (PET) degradation	Not specified	*Pseudomonas aestusnigri* VGXO14	Bollinger et al. (2020)
9	Polyvinyl chloride (PVC) degradation	Marine samples were collected at a depth of 0–30 cm of the water column from Diu Island, India	*Vibrio, Altermonas,* and *Cobetia* spp.	Khandare, Chaudhary, and Jha (2021)
10	PET and LDPE degradation	Marine waters of the Bay of Bengal, Sunderbans, West Bengal	*Vibrio* spp. (GenBank accession No.: KY941137.1 strain PD6)	Sarkhel et al. (2019)

3.3 MARINE CYANOBACTERIA

Cyanobacteria, also known as blue-green algae, are believed to be among the Earth's oldest organisms, inhabiting our planet for billions of years. In general, cyanobacteria are photoautotrophic and are capable of thriving in diverse environments, such as freshwater, marine, and soil, including extreme conditions with high salinity as well as high and low temperatures. Despite sharing common origins and basic anatomical features with bacteria, cyanobacteria exhibit different biological, ecological, and physical characteristics. Their ability to harness solar energy and perform photosynthesis through chlorophyll *a*, fixing CO_2 and producing O_2, distinguishes them as the largest photosynthetic prokaryotes (Perera et al. 2023).

In marine environments, cyanobacteria are divided into two main ecological groups: benthic and planktonic cyanobacteria (Fattom and Shilo 1984). Planktonic cyanobacteria float freely in water columns, regulating their depth using buoyancy mechanisms and gas vesicles. Prominent genera in marine planktonic communities include *Prochlorococcus*, *Synechococcus*, *Synechocystis*, and *Cynobium*. Furthermore, *Prochlorococcus* and *Synechococcus* are the most prolific photosynthetic organisms on our planet (Flombaum et al. 2013; Munoz-Marin et al. 2020). Together, *Prochlorococcus* and *Synechococcus* account for approximately 25% of the primary production that occurs within the oceans (Flombaum et al. 2013). Benthic cyanobacteria possess unique gliding motility that facilitates their interaction with solid surfaces such as sediments, algae, rocks, and stones, and aquatic plants (Fattom and Shilo 1984). The genus *Lyngbya* is one of the common components of benthic communities, forming dense and widespread cyanobacterial mats. Furthermore, certain cyanotoxins, specifically categorized as dermatotoxins, aplysiatoxins, and lyngbiatoxins, have been identified in benthic marine cyanobacteria and these toxins are known to cause severe cases of contact dermatitis in individuals who come into contact with coastal waters where these cyanobacteria are present (Zanchett and Oliveira-Filho 2013).

Cyanobacteria have evolved survival mechanisms that enable them to thrive in these challenging environments over extended periods. These mechanisms involve the production of secondary metabolites with various properties: antioxidants, photoprotectants, moisturizers, allelopathic agents, and toxins (Fidor, Konkel and Mazur-Marzec 2019). These bioactive compounds, including polysaccharides, pigments, fatty acids, and peptides found in cyanobacteria, hold significant potential for medicinal drug development (Hachicha et al. 2022; Khalifa et al. 2021). Specifically, bioactive compounds isolated from cyanobacteria encompass a wide range of activities, including antibacterial, anti-inflammatory, anti-obesity, antiparasitic, anticancer, antidiabetic, antiviral, antioxidant, anti-aging, hepatoprotective, immunomodulatory, photoprotective, and neuroprotective effects (Perera et al. 2023). To date, over 2,000 secondary metabolites have been identified from cyanobacteria, including species like *Moorea*, *Lyngbya*, and *Okeania* spp.

Beyond pharmaceuticals, cyanobacteria are found to possess diverse applications: food additives, animal feed, cosmetics, biofuel production, wastewater treatment,

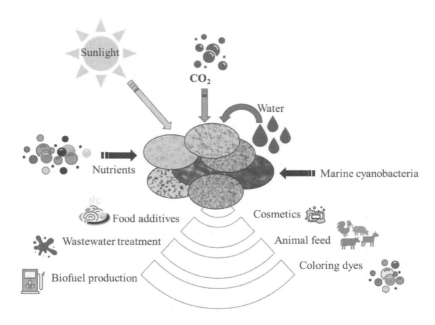

FIGURE 3.5 Potential applications of marine cyanobacteria.

coloring dyes, and biopolymer production (Figure 3.5) (Hachicha et al. 2022; Khalifa et al. 2021). Moreover, cyanobacterial pigments are valued as natural colorants and dyes for food and other applications (Nawaz et al. 2023). Additionally, cyanobacteria have demonstrated effectiveness in removing heavy metals from industrial wastewater and participating in the bioremediation of aquatic habitats (*Limnococcus* sp., *Nostocmuscorum*, and *Synechococcus* sp. PCC 7942) (Al-Amin et al. 2021). These different microorganisms have numerous applications and can be harnessed for various industrial and environmental purposes (Nawaz et al. 2023). Marine cyanobacteria like *Cylindrospermum* sp., *Phormidium* sp., *Pseudanabaena* sp., and *Spirulina* sp. have proven effective in wastewater bioremediation (Ameen et al. 2021; Haberle et al. 2020), and some are utilized in animal feed production, including *Arthrospira maxima*, *Spirulina platensis*, and *Schizochytrium* sp. (Mahata et al. 2022).

3.4 FUTURE PROSPECTIVES

Marine microorganisms are exceptionally promising and have the potential to play a pivotal role in various fields. Biotechnology is expected to advance, leading to the discovery of novel enzymes, bioactive compounds, and genes with a wide range of applications. The search for new drugs, particularly antibiotics and antiviral agents, will likely intensify, addressing the growing concern of antibiotic resistance. Marine microorganisms will continue to be valuable assets in environmental monitoring, serving as bioindicators to detect and respond to changes in marine ecosystems. Additionally, marine microbes will play a vital role in bioremediation efforts, aiding in the cleanup of polluted marine environments. These microorganisms are integral in climate change mitigation due to their carbon sequestration capabilities. Advanced genetic and molecular techniques will deepen our understanding of marine microbial ecology and foster better conservation practices. Moreover, marine microorganisms, particularly extremophiles, may find relevance in astrobiology, aiding in the search for life on other planets. This expanding knowledge of marine microorganisms will not only open doors to scientific discoveries but also contribute to environmental protection and sustainable innovation across various industries.

REFERENCES

Ahmad, A., R. Hamid, A. C. Dada, and G. Usup. 2013. "*Pseudomonas putida* strain FStm2 isolated from shark skin: A potential source of bacteriocin." *Probiotics and Antimicrobial Proteins* 5 (3):165–75. doi: 10.1007/s12602-013-9140-4

Al-Amin, A., F. Parvin, J. Chakraborty, and Y.-I. Kim. 2021. "Cyanobacteria mediated heavy metal removal: A review on mechanism, biosynthesis, and removal capability." *Environmental Technology Reviews* 10 (1):44–57. doi: 10.1080/21622515.2020.1869323

Ameen, F., A. A. Al-Homaidan, K. Alsamhary, N. M. Al-Enazi, and S. AlNadhari. 2021. "Bioremediation of ossein effluents using the filamentous marine cyanobacterium *Cylindrospermum stagnale*." *Environmental Pollution* 284:117507. doi: 10.1016/j. envpol.2021.117507

Ameen, F., S. AlNadhari, and A. A. Al-Homaidan. 2021. "Marine microorganisms as an untapped source of bioactive compounds." *Saudi Journal of Biological Sciences* 28 (1):224–231. doi: 10.1016/j.sjbs.2020.09.052

Ameen, F. A., A. M. Hamdan, and M. Y. El-Naggar. 2020. "Assessment of the heavy metal bioremediation efficiency of the novel marine lactic acid bacterium, *Lactobacillus plantarum* MF042018." *Scientific Reports* 10 (1):314. doi: 10.1038/s41598-019-57210-3

Anastasopoulou, A., and T. Fortibuoni. 2019. "Impact of Plastic Pollution on Marine Life in the Mediterranean Sea." In *Plastics in the Aquatic Environment: Part I*, edited by F. Stock, G. Reifferscheid, N. Brennholt, and E. Kostianaia, 135–196. Cham: Springer International Publishing.

Balakrishnan, S., N. Arunagirinathan, M. R. Rameshkumar, P. Indu, N. Vijaykanth, K. S. Almaary, S. M. Almutairi, and T.-W. Chen. 2022. "Molecular characterization of biosurfactant producing marine bacterium isolated from hydrocarbon-contaminated soil using 16S rRNA gene sequencing." *Journal of King Saud University: Science* 34 (3):101871. doi: 10.1016/j.jksus.2022.101871

Banerjee, G., and A. K. Ray. 2017. "The advancement of probiotics research and its application in fish farming industries." *Research in Veterinary Science* 115:66–77. doi: 10.1016/j.rvsc.2017.01.016

Barzkar, N., and M. Sohail. 2020. "An overview on marine cellulolytic enzymes and their potential applications." *Applied Microbiology and Biotechnology* 104 (16):6873–6892. doi: 10.1007/s00253-020-10692-y

Bhattacharya, S., S. Bhattacharya, R. Gachhui, S. Hazra, and J. Mukherjee. 2019. "U32 collagenase from *Pseudoalteromonas agarivorans* NW4327: Activity, structure, substrate interactions and molecular dynamics simulations." *International Journal of Biological Macromolecules* 124:635–650. doi: 10.1016/j.ijbiomac.2018.11.206

Bollinger, A., S. Thies, E. Knieps-Grunhagen, C. Gertzen, S. Kobus, A. Hoppner, M. Ferrer, H. Gohlke, S. H. J. Smits, and K. E. Jaeger. 2020. "A novel polyester hydrolase from the marine bacterium *Pseudomonas aestusnigri*: Structural and functional insights." *Frontiers in Microbiology* 11:114. doi: 10.3389/fmicb.2020.00114

Cabello, F. C. 2006. "Heavy use of prophylactic antibiotics in aquaculture: A growing problem for human and animal health and for the environment." *Environmental Microbiology* 8 (7):1137–1144. doi: 10.1111/j.1462-2920.2006.01054.x

Cabral Pinto, M. M. S., M. M. V. Silva, E. A. Ferreira da Silva, and A. P. Marinho-Reis. 2017. "The cancer and non-cancer risk of *Santiago Island* (Cape verde) population due to potential toxic elements exposure from soils." *Geosciences* 7 (3):78. doi: 10.3390/geosciences7030078

Chakraborty, S., S. Jana, A. Gandhi, K. K. Sen, W. Zhiang, and C. Kokare. 2014. "Gellan gum microspheres containing a novel alpha-amylase from marine *Nocardiopsis* sp. Strain B2 for immobilization." *International Journal of Biological Macromolecules* 70:292–299. doi: 10.1016/j.ijbiomac.2014.06.046

Chaprao, M. J., Rcfs da Silva, R. D. Rufino, J. M. Luna, V. A. Santos, and L. A. Sarubbo. 2018. "Production of a biosurfactant from *Bacillus methylotrophicus* UCP1616 for use in the bioremediation of oil-contaminated environments." *Ecotoxicology* 27 (10):1310–1322. doi: 10.1007/s10646-018-1982-9

Crisafi, F., L. Giuliano, M. M. Yakimov, M. Azzaro, and R. Denaro. 2016. "Isolation and degradation potential of a cold-adapted oil/PAH-degrading marine bacterial consortium from Kongsfjorden (Arctic region)." *Rendiconti Lincei* 27 (S1):261–270. doi: 10.1007/s12210-016-0550-6

Danso, D., J. Chow, and W. R. Streit. 2019a. "Plastics: Environmental and biotechnological perspectives on microbial degradation." *Applied and Environmental Microbiology* 85 (19):e01095-19. doi: 10.1128/AEM.01095-19

Dash, H. R., N. Mangwani, J. Chakraborty, S. Kumari, and S Das. 2013. "Marine bacteria: Potential candidates for enhanced bioremediation." *Applied Microbiology and Biotechnology* 97 (2):561–71. doi: 10.1007/s00253-012-4584-0

De, J., N. Ramaiah, and L. Vardanyan. 2008. "Detoxification of toxic heavy metals by marine bacteria highly resistant to mercury." *Marine Biotechnology (New York)* 10 (4):471–7. doi: 10.1007/s10126-008-9083-z

Debnath, M., A. K. Paul, and P. S. Bisen. 2007. "Natural bioactive compounds and biotechnological potential of marine bacteria." *Current Pharmaceutical Biotechnology* 8 (5):253–60. doi: 10.2174/138920107782109976

Dell'Anno, F., C. Brunet, L. J. van Zyl, M. Trindade, P. N. Golyshin, A. Dell'Anno, A. Ianora, and C. Sansone. 2020. "Degradation of hydrocarbons and heavy metal reduction by marine bacteria in highly contaminated sediments." *Microorganisms* 8 (9) :1402. doi: 10.3390/microorganisms8091402

DeLong, E. F., N. R. Pace, and M. Kane. 2001. "Environmental diversity of bacteria and archaea." *Systematic Biology* 50 (4):470–478.

FAO. 2020. The State of World Fisheries and Aquaculture 2020, Sustainability in action, Rome (2020).

Farhadian, S., A. Asoodeh, and M. Lagzian. 2015. "Purification, biochemical characterization and structural modeling of a potential htrA-like serine protease from *Bacillus subtilis* DR8806." *Journal of Molecular Catalysis B: Enzymatic* 115:51–58. doi: 10.1016/j.molcatb.2015.02.001

Fattom, A., and M. Shilo. 1984. "Hydrophobicity as an adhesion mechanism of benthic cyanobacteria." *Applied and Environmental Microbiology* 47 (1):135–143. doi: 10.1128/aem.47.1.135-143.1984

Fidor, A., R. Konkel, and H. Mazur-Marzec. 2019. "Bioactive peptides produced by cyanobacteria of the genus *Nostoc*: A review." *Marine Drugs* 17 (10):561. doi: 10.3390/md17100561

Flombaum, P., J. L. Gallegos, R. A. Gordillo, J. Rincon, L. L. Zabala, N. Jiao, D. M. Karl, W. K. Li, M. W. Lomas, D. Veneziano, C. S. Vera, J. A. Vrugt, and A. C. Martiny. 2013. "Present and future global distributions of the marine cyanobacteria

Prochlorococcus and *Synechococcus.*" *Proceedings of the National Academy of Sciences of the United States of America* 110 (24):9824–9829. doi: 10.1073/pnas.1307701110

Fulke, A. B., A. Kotian, and M. D. Giripunje. 2020. "Marine microbial response to heavy metals: Mechanism, implications and future prospect." *Bulletin of Environmental Contamination and Toxicology* 105 (2):182–197. doi: 10.1007/s00128-020-02923-9

Gaber, Y., S. Mekasha, G. Vaaje-Kolstad, V. G. H. Eijsink, and M. W. Fraaije. 2016. "Characterization of a chitinase from the cellulolytic actinomycete *Thermobifida fusca.*" *Biochimica et Biophysica Acta* 1864 (9):1253–1259. doi: 10.1016/j.bbapap.2016.04.010

Geyer, R., J. R. Jambeck, and K. L. Law. 2017. "Production, use, and fate of all plastics ever made." *Science Advances* 3 (7):e1700782. doi: 10.1126/sciadv.1700782

Ghattavi, S., and A. Homaei. 2023. "Marine enzymes: Classification and application in various industries." *International Journal of Biological Macromolecules* 230:123136. doi: 10.1016/j.ijbiomac.2023.123136

Ghosh, S., T. Sarkar, S. Pati, Z. A. Kari, H. A. Edinur, and R. Chakraborty. 2022. "Novel bioactive compounds from marine sources as a tool for functional food development." *Frontiers in Marine Science* 9. doi: 10.3389/fmars.2022.832957

Gundogan, N. 2014. "Klebsiella." In *Encyclopedia of Food Microbiology*, edited by C. A. Batt and M. L. Tortorello, 383–388. Oxford: Academic Press.

Guo, Y., Z. Gao, J. Xu, S. Chang, B. Wu, and B. He. 2018. "A family 30 glucurono-xylanase from *Bacillus subtilis* LC9: Expression, characterization and its application in Chinese bread making." *International Journal of Biological Macromolecules* 117:377–384. doi: 10.1016/j.ijbiomac.2018.05.143

Haberle, I., E. Hrustić, I. Petrić, E. Pritišanac, T. Šilović, L. Magić, S. Geček, A. Budiša, and M. Blažina. 2020. "Adriatic cyanobacteria potential for cogeneration biofuel production with oil refinery wastewater remediation." *Algal Research* 50:101978. doi: 10.1016/j.algal.2020.101978

Hachicha, R., F. Elleuch, H. Ben Hlima, P. Dubessay, H. de Baynast, C. Delattre, G. Pierre, R. Hachicha, S. Abdelkafi, P. Michaud, and I. Fendri. 2022. "Biomolecules from microalgae and cyanobacteria: Applications and market survey." *Applied Sciences* 12 (4):1924. doi: 10.3390/app12041924

Hamdan, A. M., H. Abd-El-Mageed, and N. Ghanem. 2021. "Biological treatment of hazardous heavy metals by *Streptomyces rochei* ANH for sustainable water management in agriculture." *Science Reports* 11 (1):9314. doi: 10.1038/s41598-021-88843-y

Hamidi, M., P. S. Kozani, P. S. Kozani, G. Pierre, P. Michaud, and C. Delattre. 2019. "Marine bacteria versus microalgae: Who is the best for biotechnological production of bioactive compounds with antioxidant properties and other biological applications?" *Marine Drugs* 18 (1). doi: 10.3390/md18010028

Hassan, S., Z. Sabreena, S Khurshid, A. Bhat, V. Kumar, F. Ameen, and A. Ganai. 2022. "Marine bacteria and omic approaches: A novel and potential repository for bioremediation assessment." *Journal of Applied Microbiology* 133 (4):2299–2313. doi: 10.1111/jam.15711

Hohn, S., E. Acevedo-Trejos, J. F. Abrams, J. Fulgencio de Moura, R. Spranz, and A. Merico. 2020. "The long-term legacy of plastic mass production." *The Science of the Total Environment* 746:141115. doi: 10.1016/j.scitotenv.2020.141115

Huang, L., T. Ma, D. Li, F. L. Liang, R. L. Liu, and G. Q Li. 2008. "Optimization of nutrient component for diesel oil degradation by *Rhodococcus erythropolis.*" *Marine Pollution Bulletin* 56 (10):1714–8. doi: 10.1016/j.marpolbul.2008.07.007

Ilhami, M. A., and H. Holifah. 2022. "Marine bioremediation using *Alcanivorax borkumensis* SK2 as a waste prevention oil industry on the Tunda Island to the impact of flood in Banten region." *GMPI Conference Series* 1:92–103. doi: 10.53889/gmpics.v1.89

Joshi, G., P. Goswami, P. Verma, G. Prakash, P. Simon, N. V. Vinithkumar, and G. Dharani. 2022. "Unraveling the plastic degradation potentials of the plastisphere-associated marine bacterial consortium as a key player for the low-density polyethylene degradation." *Journal of Hazardous Materials* 425:128005. doi: 10.1016/j.jhazmat.2021.128005

Kar, D., P. Sur, S. K. Mandai, T. Saha, and R. K. Kole. 2007. "Assessment of heavy metal pollution in surface water." *International Journal of Environmental Science & Technology* 5 (1):119–124. doi: 10.1007/bf03326004

Khalifa, S. A. M., E. S. Shedid, E. M. Saied, A. R. Jassbi, F. H. Jamebozorgi, M. E. Rateb, M. Du, M. M. Abdel-Daim, G.-Y. Kai, M. A. M. Al-Hammady, J. Xiao, Z. Guo, and H. R. El-Seedi. 2021. "Cyanobacteria: From the oceans to the potential biotechnological and biomedical applications." *Marine Drugs* 19 (5):241. doi: 10.3390/md19050241

Khandare, S. D., D. R. Chaudhary, and B. Jha. 2021. "Bioremediation of polyvinyl chloride (PVC) films by marine bacteria." *Marine Pollution Bulletin* 169:112566. doi: 10.1016/j.marpolbul.2021.112566

Kubicki, S., A. Bollinger, N. Katzke, K. E. Jaeger, A. Loeschcke, and S. Thies. 2019. "Marine biosurfactants: Biosynthesis, structural diversity and biotechnological applications." *Marine Drugs* 17 (7). doi: 10.3390/md17070408

Kumari, A., D. R. Chaudhary, and B. Jha. 2019. "Destabilization of polyethylene and polyvinylchloride structure by marine bacterial strain." *Environmental Science and Pollution Research International* 26 (2):1507–1516. doi: 10.1007/s11356-018-3465-1

Kumari, S., and S. Das. 2019. "Expression of metallothionein encoding gene bmtA in biofilm-forming marine bacterium *Pseudomonas aeruginosa* N6P6 and understanding its involvement in pb(II) resistance and bioremediation." *Environmental Science and Pollution Research International* 26 (28):28763–28774. doi: 10.1007/s11356-019-05916-2

Lee, D. W., H. Lee, B. O. Kwon, J. S. Khim, U. H. Yim, B. S. Kim, and J. J Kim. 2018. "Biosurfactant-assisted bioremediation of crude oil by indigenous bacteria isolated from Taean beach sediment." *Environmental Pollution* 241:254–264. doi: 10.1016/j.envpol.2018.05.070

Liu, K., T. Wu, X. Wang, Z. Song, C. Zong, N. Wei, and D Li. 2019. "Consistent transport of terrestrial microplastics to the ocean through atmosphere." *Environmental Science & Technology* 53 (18):10612–10619. doi: 10.1021/acs.est.9b03427

Liu, R., S. Zhao, B. Zhang, G. Li, X. Fu, P. Yan, and Z. Shao. 2023. "Biodegradation of polystyrene (PS) by marine bacteria in mangrove ecosystem." *Journal of Hazardous Materials* 442:130056. doi: 10.1016/j.jhazmat.2022.130056

Ma, C., X. Lu, C. Shi, J. Li, Y. Gu, Y. Ma, Y. Chu, F. Han, Q. Gong, and W Yu. 2007. "Molecular cloning and characterization of a novel beta-agarase, AgaB, from marine *Pseudoalteromonas* sp. CY24." *The Journal of Biological Chemistry* 282 (6): 3747–3754. doi: 10.1074/jbc.M607888200

Mahata, C., P. Das, S. Khan, M. I. A. Thaher, M. Abdul Quadir, S. N. Annamalai, and H. Al Jabri. 2022. "The potential of marine microalgae for the production of food, feed, and fuel (3F)." *Fermentation* 8 (7). doi: 10.3390/fermentation8070316

Martinez Cruz, P., A. L. Ibanez, O. A. Monroy Hermosillo, and H. C. R. Saad. 2012. "Use of probiotics in aquaculture." *ISRN Microbiology* 2012:916845.

Mishra, M., S. K. Singh, and A. Kumar. 2021. "Role of Omics Approaches in Microbial Bioremediation." In *Microbe Mediated Remediation of Environmental Contaminants*, edited by A. Kumar, V. K. Singh, P. Singh, and V. K. Mishra, 435–445. Woodhead Publishing.

Mishra, V. K., A. R. Upadhyaya, S. K. Pandey, and B. D. Tripathi. 2008. "Heavy metal pollution induced due to coal mining effluent on surrounding aquatic ecosystem and its management through naturally occurring aquatic macrophytes." *Bioresource Technology* 99 (5):930–936. doi: 10.1016/j.biortech.2007.03.010

Mohanrasu, K., N. Premnath, G. Siva Prakash, M. Sudhakar, T. Boobalan, and A. Arun. 2018. "Exploring multi potential uses of marine bacteria: An integrated approach for PHB production, PAHs and polyethylene biodegradation." *Journal of Photochemistry and Photobiology* 185:55–65. doi: 10.1016/j.jphotobiol.2018.05.014

Mora-Teddy, A. K., and C. D. Matthaei. 2019. "Microplastic pollution in urban streams across New Zealand: Concentrations, composition and implications." *New Zealand Journal of Marine and Freshwater Research* 54 (2):233–250. doi: 10.1080/00288330.2019.1703015

Mounier, J., A. Camus, I. Mitteau, P.-J. Vaysse, P. Goulas, R. Grimaud, and P. Sivadon. 2014. "The marine bacterium *Marinobacter hydrocarbonoclasticus* SP17 degrades a wide range of lipids and hydrocarbons through the formation of oleolytic biofilms with distinct gene expression profiles." *FEMS Microbiology Ecology* 90 (3):816–831. doi: 10.1111/1574-6941.12439

Munoz-Marin, M. C., G. Gomez-Baena, A. Lopez-Lozano, J. A. Moreno-Cabezuelo, J. Diez, and J. M. Garcia-Fernandez. 2020. "Mixotrophy in marine picocyanobacteria: Use of organic compounds by *Prochlorococcus* and *Synechococcus*." *ISME Journal* 14 (5):1065–1073. doi: 10.1038/s41396-020-0603-9

Napper, I. E., and R. C. Thompson. 2019. "Marine Plastic Pollution: Other Than Microplastic." In *Waste*, edited by T. M. Letcher and D. A. Vallero, 425–442. Academic Press.

Nawaz, T., L. Gu, S. Fahad, S. Saud, Z. Jiang, S. Hassan, M. T. Harrison, K. Liu, M. Ahmad Khan, H. Liu, K. El-Kahtany, C. Wu, M. Zhu, and R. Zhou. 2023. "A comprehensive review of the therapeutic potential of cyanobacterial marine bioactives: Unveiling the hidden treasures of the sea." *Food and Energy Security* 12 (5):e495. doi: 10.1002/fes3.495

Orekhova, N. A., Y. A. Davydova, and G. Y. Smirnov. 2023. "Structural-functional aberrations of erythrocytes in the northern red-backed vole (*Clethrionomys rutilus* Pallas, 1779) that inhabits the zone of influence of the copper smelter (the Middle Ural)." *Biometals* 36 (4):847–864. doi: 10.1007/s10534-022-00478-2

Pang, G., X. Li, M. Ding, S. Jiang, P. Chen, Z. Zhao, R. Gao, B. Song, X. Xu, Q. Shen, F. M. Cai, and I. S. Druzhinina. 2023. "The distinct plastisphere microbiome in the terrestrial-marine ecotone is a reservoir for putative degraders of petroleum-based polymers." *Journal of Hazardous Materials* 453:131399. doi: 10.1016/j.jhazmat.2023.131399

Peng, F., Z. Liu, L. Wang, and Z. Shao. 2007. "An oil-degrading bacterium: *Rhodococcus erythropolis* strain 3C-9 and its biosurfactants." *Journal of Applied Microbiology* 102 (6):1603–1611. doi: 10.1111/j.1365-2672.2006.03267.x

Pereira, E., A. P. Napp, S. Allebrandt, R. Barbosa, J. Reuwsaat, W. Lopes, L. Kmetzsch, C. C. Staats, A. Schrank, A. Dallegrave, M. do Carmo R. Peralba, L. M. P. Passaglia, F. M. Bento, and M. H. Vainstein. 2019. "Biodegradation of aliphatic and polycyclic aromatic hydrocarbons in seawater by autochthonous microorganisms." *International Biodeterioration & Biodegradation* 145:104789. doi: 10.1016/j.ibiod.2019.104789

Perera, R. M. T. D., K. H. I. N. M. Herath, K. K. A. Sanjeewa, and T. U. Jayawardena. 2023. "Recent reports on bioactive compounds from marine cyanobacteria in relation to human health applications." *Life* 13 (6):1411. doi: 10.3390/life13061411

Prieto, M. L., L. O'Sullivan, S. P. Tan, P. McLoughlin, H. Hughes, P. M. O'Connor, P. D. Cotter, P. G. Lawlor, and G. E. Gardiner. 2012. "Assessment of the bacteriocinogenic potential of marine bacteria reveals lichenicidin production by seaweed-derived *Bacillus* spp." *Marine Drugs* 10 (10):2280–2299. doi: 10.3390/md10102280

Priyadarshanee, M., and S. Das. 2021. "Bioremediation potential of biofilm forming multi-metal resistant marine bacterium *Pseudomonas chengduensis* PPSS-4 isolated from contaminated site of Paradip Port, Odisha." *Journal of Earth System Science* 130 (3):125. doi: 10.1007/s12040-021-01627-w

Qasem, N. A. A., R. H. Mohammed, and D. U. Lawal. 2021. "Removal of heavy metal ions from wastewater: A comprehensive and critical review." *NPJ Clean Water* 4 (1):36. doi: 10.1038/s41545-021-00127-0

Rather, I. A., R. Galope, V. K. Bajpai, J. Lim, W. K. Paek, and Y.-H. Park. 2017. "Diversity of marine bacteria and their bacteriocins: Applications in aquaculture." *Reviews in Fisheries Science & Aquaculture* 25 (4):257–269. doi: 10.1080/23308249.2017.1282417

Roager, L., and E. C. Sonnenschein. 2019. "Bacterial candidates for colonization and degradation of marine plastic debris." *Environmental Science & Technology* 53 (20): 11636–11643. doi: 10.1021/acs.est.9b02212

Sangeetha Devi, R., R. Ramya, K. Kannan, A. Robert Antony, and V. R. Kannan. 2019. "Investigation of biodegradation potentials of high density polyethylene degrading marine bacteria isolated from the coastal regions of Tamil Nadu, India." *Marine Pollution Bulletin* 138:549–560. doi: 10.1016/j.marpolbul.2018.12.001

Sarkhel, R., S. Sengupta, P. Das, and A. Bhowal. 2019. "Comparative biodegradation study of polymer from plastic bottle waste using novel isolated bacteria and fungi from marine source." *Journal of Polymer Research* 27 (1):16. doi: 10.1007/s10965-019-1973-4

Sathishkumar, R., G. Ananthan, K. Iyappan, and C. Stalin. 2015. "A statistical approach for optimization of alkaline lipase production by ascidian associated–*Halobacillus trueperi* RSK CAS9." *Biotechnology Reports* 8:64–71. doi: 10.1016/j.btre.2015.09.002

Seagren, E. A. 2023. "Bioremediation." *Reference Module in Biomedical Sciences*. Elsevier.

Sequeiros, C., M. E. Garces, M. Vallejo, E. R. Marguet, and N. L. Olivera. 2015. "Potential aquaculture *Probiont lactococcus* lactis TW34 produces nisin Z and inhibits the fish pathogen *Lactococcus garvieae*." *Archives of Microbiology* 197 (3):449–458. doi: 10.1007/s00203-014-1076-x

Sharma, S., and S. Chatterjee. 2017. "Microplastic pollution, a threat to marine ecosystem and human health: A short review." *Environmental Science and Pollution Research International* 24 (27):21530–21547. doi: 10.1007/s11356-017-9910-8

Shrestha, R., S. Ban, S. Devkota, S. Sharma, R. Joshi, A. P. Tiwari, H. Y. Kim, and M. K. Joshi. 2021. "Technological trends in heavy metals removal from industrial wastewater: A review." *Journal of Environmental Chemical Engineering* 9 (4):105688. doi: 10.1016/j.jece.2021.105688

Singh, N., U. Goutam, and M. Ghosh. 2022. "Deep-Marine Bacteria: The Frontier Alternative for Heavy Metals Bioremediation." In *Development in Wastewater Treatment Research and Processes*, edited by S. Rodriguez-Couto and M. P. Shah, 429–450. Elsevier.

Sivaperumal, P., K. Kamala, and R. Rajaram. 2017. "Chapter Eight: Bioremediation of Industrial Waste through Enzyme Producing Marine Microorganisms." In *Advances in Food and Nutrition Research*, edited by S.-K. Kim and F. Toldrá, 165–179. Academic Press.

Statista Inc. 2022. "Annual production of plastics worldwide from 1950 to 2021." https://www.statista.com/statistics/282732/global-production-of-plastics-since-1950/ (accessed August 20, 2023).

Toshchakov, S. V., A. A. Korzhenkov, T. N. Chernikova, M. Ferrer, O. V. Golyshina, M. M. Yakimov, and P. N Golyshin. 2017. "The genome analysis of *Oleiphilus messinensis* ME102 (DSM 13489(T)) reveals backgrounds of its obligate alkane-devouring marine lifestyle." *Marine Genomics* 36:41–47. doi: 10.1016/j.margen.2017.07.005

Tuuri, E. M., and S. C. Leterme. 2023. "How plastic debris and associated chemicals impact the marine food web: A review." *Environmental Pollution* 321:121156. doi: 10.1016/j.envpol.2023.121156

Vardhan, K. H., P. S. Kumar, and R. C. Panda. 2019. "A review on heavy metal pollution, toxicity and remedial measures: Current trends and future perspectives." *Journal of Molecular Liquids* 290:111197. doi: 10.1016/j.molliq.2019.111197

Verma, R., K. S. Vinoda, M. Papireddy, and A. N. S. Gowda. 2016. "Toxic pollutants from plastic waste: A review." *Procedia Environmental Sciences* 35:701–708. doi: 10.1016/j.proenv.2016.07.069

Wagner-Döbler, I., W. Beil, S. Lang, M. Meiners, and H. Laatsch. 2002. "Integrated Approach to Explore the Potential of Marine Microorganisms for the Production of Bioactive Metabolites." In *Tools and Applications of Biochemical Engineering Science*, edited by K. Schügerl et al., 207–238. Berlin: Springer.

Wang, F., M. Li, L. Huang, and X. H. Zhang. 2021. "Cultivation of uncultured marine microorganisms." *Marine Life Science and Technology* 3 (2):117–120. doi: 10.1007/s42995-021-00093-z

Watanabe, K. 2001. "Microorganisms relevant to bioremediation." *Current Opinion in Biotechnology* 12 (3):237–241. doi: 10.1016/s0958-1669(00)00205-6

Yusuf, A. A., J. D. Ampah, M. E. M. Soudagar, I. Veza, U. Kingsley, S. Afrane, C. Jin, H. Liu, A. Elfasakhany, and K. A. Buyondo. 2022. "Effects of hybrid nanoparticle additives in *n*-butanol/waste plastic oil/diesel blends on combustion, particulate and gaseous emissions from diesel engine evaluated with entropy-weighted PROMETHEE II and TOPSIS: Environmental and health risks of plastic waste." *Energy Conversion and Management* 264:115758. doi: 10.1016/j.enconman.2022.115758

Zadjelovic, V., G. Erni-Cassola, T. Obrador-Viel, D. Lester, Y. Eley, M. I. Gibson, C. Dorador, P. N. Golyshin, S. Black, E. M. H. Wellington, and J. A. Christie-Oleza. 2022. "A mechanistic understanding of polyethylene biodegradation by the marine bacterium *Alcanivorax*." *Journal of Hazardous Materials* 436:129278. doi: 10.1016/j.jhazmat.2022.129278

Zamora-Ledezma, C., D. Negrete-Bolagay, F. Figueroa, E. Zamora-Ledezma, M. Ni, F. Alexis, and V. H. Guerrero. 2021. "Heavy metal water pollution: A fresh look about hazards, novel and conventional remediation methods." *Environmental Technology & Innovation* 22:101504. doi: 10.1016/j.eti.2021.101504

Zanchett, G., and E. C. Oliveira-Filho. 2013. "Cyanobacteria and cyanotoxins: From impacts on aquatic ecosystems and human health to anticarcinogenic effects." *Toxins* 5 (10):1896–1917. doi: 10.3390/toxins5101896

Zhang, C., and S.-K. Kim. 2012. "Chapter 28: Application of Marine Microbial Enzymes in the Food and Pharmaceutical Industries." In *Advances in Food and Nutrition Research*, edited by S.-K. Kim, 423–435. Academic Press.

Zhu, B., and L. Ning. 2016. "Purification and characterization of a new kappa-carrageenase from the marine bacterium *Vibrio* sp. NJ-2." *Journal of Microbiolog and Biotechnology* 26 (2):255–62. doi: 10.4014/jmb.1507.07052

4 Marine Fungi

4.1 GENERAL FACTS ABOUT FUNGI

Marine fungi represent an ecological grouping rather than a taxonomic one, encompassing around 1,800 fungi species, excluding those that form lichens. Habitats of marine fungi are reported from most tested marine ecosystems and generally have a pantropical or pan-temperate distribution (Hyde et al. 1998). Fungi play a vital role in terrestrial and freshwater ecosystems by helping to recycle plant debris and participating in elemental cycles (Ekblad et al. 2013; Gadd 2007). Marine fungi release CO_2 into the atmosphere and support to increase fertility of soil via supporting to release nitrogen and phosphorus from decomposing materials (Gadd 2007). However, in the marine environment, fungi are associated with debris such as driftwood, living organisms, or seafloor sediments and are only recognized to play a key role in the element cycle in deep-sea sediments (Baltar, Zhao and Herndl 2021; Bengtson et al. 2017). Furthermore, marine fungi serve a significant role as pathogens affecting marine plants and animals. Additionally, they engage in symbiotic relationships with various other organisms in marine ecosystems (Agrawal et al. 2018). Specifically, harsh chemical and physical conditions in marine ecosystems lead to the production of diverse secondary metabolites by marine fungi. According to previous observations, some fungi species' habitats in marine ecosystems have developed specific metabolic pathways which have not been reported in terrestrial fungi species so far (Abdel-Lateff 2008). Thus taken together, marine fungi represent a promising source to extract natural bioactive secondary metabolite for numerous industries, such as pharmaceuticals, cosmetics, nutraceuticals, and agrochemicals.

4.2 CLASSIFICATIONS, ORIGIN, AND HABITATS OF MARINE FUNGI

Throughout time, various definitions of marine fungi have emerged. Initially, marine fungi were categorized based on their physiological traits, notably their need for a salinity level exceeding 30% to facilitate growth (Jennings 1983). Even at present, it is difficult to give a specific definition of a marine fungus to the diverse range of habitats of marine fungi. However, the most well-known and frequently cited definition of 'marine fungi' was introduced by Kohlmeyer and Kohlmeyer back in 1979. According to Kohlmeyer and Kohlmeyer, marine fungi can be classified into two main categories as obligate and facultative (Kohlmeyer and Kohlmeyer 1979). Obligate marine fungi exclusively thrive and produce spores in marine or estuary ecosystems. On the other hand, facultative marine fungi have the ability to grow in both freshwater and marine environments. Marine fungi exhibit diverse lifestyles (Calado Mda et al. 2015; Pang et al. 2016). They can exist in obligate or facultative modes and can either float freely in the water or attach themselves to various substrates such as sand grains, marine higher plants, corals, algae, sponges, mollusks, fish, and more (Calado Mda et al. 2015; Pang et al. 2016; Raghukumar 2017).

The exact origin of fungi remains a subject of uncertainty. It is generally believed that fungi began to evolve in the late Proterozoic era, roughly between 900 and 570 million years ago (Berbee et al. 2020). However, evidence from the Ongeluk fossils (South Africa) challenges these estimates, suggesting that the fungal clade may be significantly older than the present-age estimates (2.4 billion years) of fungi. The basalt formations from the Palaeoproterozoic Ongeluk Formation in South Africa indicate that fungi might have existed around 2.4 billion years ago in association with submarine volcanoes (Bengtson et al. 2017).

In general, marine fungi belong to several phyla such as Ascomycota, Deuteromycota, Chytridiomycota, Bacidomycota, and Zygomycota (Behera and Das 2023; Kornprobst 2014). Furthermore, compared to other microorganisms such as marine bacteria, the diversity of marine fungi is not yet well documented (Jones et al. 2019). At present, the marine fungi website (www.marinefungi.org) lists a total of 2,041 species, belonging to 10 phyla, 34 classes, 108 orders, 278 families, and 814 genera (as of March 5, 2024) (Mushroom Research Foundation 2024). Interestingly, many of the species described in ocean ecosystems have close relations to fungi species reported from terrestrial ecosystems. Fungi have, in fact, transitioned multiple times between terrestrial and marine environments and vice versa. During the course of evolution in the Precambrian era (spanning from the formation of Earth around 4,600 million years ago to the onset of the Cambrian Period about 541 million years ago), cooling events and a decrease in salinity of marine ecosystems led to increased levels of dissolved oxygen. This change allowed early non-marine fungi to adapt and transition into marine ecosystems. The existence of terrestrial runoff or various migration pathways from terrestrial to marine environments prompted a number of adaptations in response to these ecosystem shifts. Moreover, marine fungi are found in a wide range of marine settings, including the deep sea and polar ice caps (Amend et al. 2019). Habitats of marine fungi are reported in marine sediments as well as most of the living and dead organic matter in marine ecosystems. The population density of marine fungi in ocean waters are low compared to the marine bacteria. In addition, the majority of research on marine fungi has focused on species connected to marine sediments, with specific substrates like corals, driftwood, seaweeds, and sponges (Richards et al. 2012).

4.3 APPLICATION OF MARINE FUNGI

4.3.1 ENZYMES ISOLATED FROM MARINE FUNGI

Over the past few decades, there has been a growing interest in the biotechnological application of metabolites derived from blue bioresources, particularly seaweeds and microalgae (Sanjeewa et al. 2023). Many of these valuable metabolites require efficient extraction processes to achieve the high yields necessary for industrial development. One promising extraction method currently in use is enzyme-assisted extraction, which can be combined with other extraction techniques (Sanjeewa et al. 2017). However, most of the commercial enzymes employed for extracting valuable compounds from marine materials are derived from terrestrial microorganisms and have limited substrate specificity (Sanjeewa et al. 2017).

To enhance extraction efficiency, it is advantageous to utilize more specific enzymes tailored to the target matrix. Marine fungi have emerged as promising candidates for producing degradation enzymes, and their significance in this regard has recently gained recognition. However, marine fungi remain less studied compared to marine bacteria. This chapter presents a state-of-the-art overview of degradation enzymes sourced from marine fungi, with a specific focus on enzymes involved in polysaccharide, protein, and lipid degradation.

4.3.2 POLYCYCLIC AROMATIC HYDROCARBONS (PAHs) DEGRADATION PROPERTIES OF MARINE FUNGI

PAHs are widely distributed in the environment and have the potential to persist for extended periods (more in Section 3.2.1.5). These PAH molecules consist of two or more fused benzene rings and most of them are formed during the incomplete combustion of organic materials such as petroleum products, wood and fossil fuels, and coal (Rengarajan et al. 2015). Natural sources of PAHs include forests, volcanic eruptions, oil seeps, and tree exudates. Anthropogenic sources encompass activities like municipal solid waste incineration, fossil fuel burning, coal tar production, wood burning, waste disposal, and petroleum spills and discharges (Ambade et al. 2022b). The environmental behavior of PAHs is influenced by their molecular weight, with low-molecular-weight (LMW) PAHs and high-molecular-weight (HMW) PAHs exhibiting different chemical properties (Ambade et al. 2022b). HMW PAHs, which comprise PAHs with four or more rings, tend to be less water soluble, less volatile, and more lipophilic compared to LMW PAHs, which consist of PAHs with three or fewer rings (Abdollahi et al. 2013; Tolosa et al. 2005). Moreover, a large number of studies have highlighted the fact that most PAHs are mutagenic, toxic, and carcinogenic to humans (Ambade et al. 2022a; da Silva Junior et al. 2021; Yamini and Rajeswari 2023).

PAHs from the unregulated discharge of terrestrial drainage contaminate the aquatic ecosystems and ultimately marine ecosystems. Prolonged and consistent discharges of PAHs pose a significant threat to aquatic organisms. Therefore, it is essential to comprehensively grasp the occurrence, distribution, and detrimental effects of PAHs on marine environments. Thus, many studies have focused on mitigating PAHs-related environmental impacts using microbial communities habitats in aquatic ecosystems.

Numerous fungi employ enzymes such as cytochrome P450, manganese peroxidase, lignin peroxidase, laccase, and epoxide hydrolase to metabolize PAHs (Alao and Adebayo 2022; Srivastava and Kumar 2019; Yamini and Rajeswari 2023). These metabolic processes result in the production of various compounds, including *trans*-dihydrodiols, epoxides, quinones, phenols, dihydrodiol, and tetraols, which are less toxic compared to the PAHs (Figure 4.1). Some of these metabolites can undergo

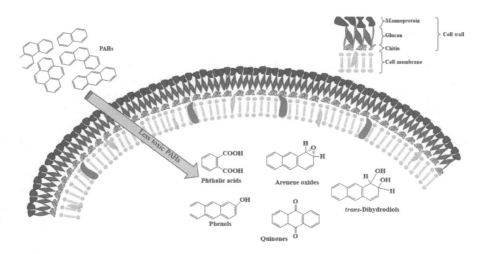

FIGURE 4.1 Breaking down of high toxic PAHs into less toxic compounds by fungi.

conjugation, forming compounds like glucuronides, glucosides, xylosides, and sulfates (Srivastava and Kumar 2019; Yamini and Rajeswari 2023).

The cultivation of fungi capable of degrading PAHs holds promise for bioremediation efforts aimed at cleaning up contaminated soils, sediments, and water sources. This approach might contribute to the reduction of PAH pollution in the environment (Marco-Urrea, Garcia-Romera and Aranda 2015). This section summarizes some potential marine fungi that can be used in PAH's degradation applications.

4.3.2.1   Case Studies

In a recent study by Birolli et al. (2018), the biodegradation potential of anthracene and various PAHs was demonstrated using the marine-derived fungus *Cladosporium* sp. CBMAI 1237. According to the authors, *Cladosporium* sp., identified as the most efficient strain, successfully biodegraded anthracene, resulting in the production of anthrone and subsequently anthraquinone. Remarkably, this strain also exhibited the ability to biotransform anthraquinone, pyrene, anthrone, fluorene, phenanthrene, acenaphthene, fluoranthene, and nitropyrene, showcasing its versatility in PAH biodegradation. In another study, Mahajan et al. (2021) attempted to screen PAH's degrading fungi from sediments collected from different locations in India. According to the authors, *Aspergillus versicolor* (NPDF190-C1-26) and *Penicillium ilerdanum* (NPDF1239-K3-F21) showed a maximum degradation of >75% for all the tested PAHs under study. Additionally, Alvarez-Barragan et al. (2021) also documented the capability of *Alternaria destruens* F10.81 in the removal of PAHs. The authors observed that *A. destruens* F10.81 achieved removal rates of PAHs exceeding 80% for phenanthrene and fluoranthene pyrene, and approximately 65% for benzo[*a*]pyrene (Alvarez-Barragan et al. 2021). Besides, it was revealed that fungi isolated from sediments at depths around

1.3–2.5 km below the seafloor exhibited varying anaerobic degradation rates for PAHs, spanning from 3% to 25%. Among these fungi, *Schizophyllium commune* 20R-7-F01, a white-rot fungus, demonstrated the highest degradation rates, reaching 25% for phenanthrene, 18% for pyrene, and 13% for benzo[*a*]pyrene after ten days of anaerobic incubation. Moreover, phenanthrene was efficiently utilized, with approximately 40.4% degradation after 20 days of anaerobic incubation (Zain Ul Arifeen et al. 2022). The fungus *Alternaria alternata* was also found to possess properties related to the colonization in polyethylene (PE) film (Gao, Liu and Sun 2022). According to the scanning electron microscope, observations revealed that *A. alternata* efficiently degraded the PE film, creating numerous visible holes on the plastic surface. Furthermore, X-ray diffraction analysis showed a significant reduction in the relative crystallinity degree of the PE film treated by strain *A. alternata* for 28 days, decreasing from 62.79% to 52.02%. Notably, the molecular weight of the PE film decreased by 95% after 120 days of treatment by strain *A. alternata*. These case studies highlight the marine fungi's potent capability to degrade and modify PAHs, potentially offering environment-friendly solutions for plastic waste management. A variety of fungal species that have been utilized for PAH degradation can be found in Table 4.1.

TABLE 4.1

Summary of PAH-Degrading Fungi Reported from Different Geographical Locations

No.	Application	Place Collected	Name of Fungi	Reference
1	Anthracene biodegradation	The marine-derived fungi were isolated from marine sponges obtained from a noncontaminated site at São Sebastião	*Cladosporium* sp. CBMAI 1237	Birolli et al. (2018)
2	Anthraquinone, anthrone, acenaphthene, fluorene, phenanthrene, fluoranthene, pyrene, and nitropyrene degradation	Gulf of Kutch and Khambhat; and from Arabian Sea	*Aspergillus versicolor* strain NPDF190-C1-26 *Penicillium ilerdanum* strain NPDF1239-K3-F21	Mahajan et al. (2021)
3	Pyrene and fluoranthene degradation	Oil-polluted sediment collected from different coastal areas (country not reported)	*lternaria destruens* F10.81	Alvarez-Barragan et al. (2021)
4	Phenanthrene, pyrene, and benzo[*a*]pyrene degradation	Anaerobic coal-associated sediments at ~1.3–2.5 km deep below the seafloor	*Schizophyllium commune*	Zain Ul Arifeen et al. (2022)
5	Polyethylene degradation	Huiquan Bay (Qingdao, China)	*Alternaria alternata*	Gao, Liu and Sun (2022)

4.3.3 HEAVY METALS BIOABSORPTION PROPERTIES OF MARINE FUNGI

According to previous studies, fungi thrive in heavy metals (cadmium, arsenic, mercury chromium, and lead)-contaminated ecosystems and are found to possess heavy metal-absorbing properties (Sharif et al. 2022). Biomolecules, including polysaccharides, chitin, phosphate, and glucuronic acid, present within the cells and cell walls of fungi are key players in the process of adsorbing heavy metals through coordination and ion exchange (Chen et al. 2014; Freitas, Roca and Reis 2015; Kumar, Kumar and Singh 2019). The presence of diverse functional groups and ionizable groups has a variable impact on their adsorption capacity (Freitas, Roca and Reis 2015; Kumar, Kumar and Singh 2019). Additionally, the specific fungal strain's affinity for the particular heavy metal ion further influences the rate of adsorption (Babich and Stotzky 1983; Kumar, Kumar and Singh 2019; Taboski, Rand and Piorko 2005). The following sections briefly discuss the potential of marine fungi to treat heavy metals.

4.3.3.1 Case Studies

Previously, several studies reported the heavy absorption properties of marine fungi. According to Taboski, Rand, and Piorko (2005), the marine fungi *Corollospora lacera* and *Monodictys pelagica* were found to absorb lead and cadmium from their culture medium. Moreover, the authors reported that approximately 93% of all the lead sequestered by *C. lacera* is located extracellularly. *M. pelagica* was found to bioaccumulate approximately 6 mg of lead and 60 mg of cadmium per 1 g of mycelium; in addition, *C. lacera* bioaccumulated around 250 mg of lead and over 7 mg of cadmium per 1 g of fungi mycelium (Taboski, Rand and Piorko 2005). In another study, Vala (2010) reported the tolerance and removal properties of arsenic by a facultative marine fungus, *Aspergillus candidus*. The authors attempted to evaluate the effects of different concentrations of trivalent and pentavalent arsenic on the growth of *A. candidus* under laboratory conditions, and the results suggest that *A. candidus* has the potential to grow in different concentrations (25 mg/L and 50 mg/L) of trivalent and pentavalent forms of arsenic (Vala 2010). The findings of this study highlight the potential use of *A. candidus* as an organism in heavy metal bioremediation applications. In another study, uranium(VI) removal from aqueous solution by poly(amic acid)-marine-derived modified mangrove endophytic fungus (*Usarium* sp. #ZZF51) was reported by Chen et al. (2014). According to the authors, compared with uranium(VI) removal of the pristine biomass, the maximum uranium(VI) adsorption capacity of the modified biomass increased 9.5-fold under the optimal conditions of pH 5.0, S/L 0.4, and an equilibrium time of 180 minutes (Chen et al. 2014).

4.3.4 SURFACE-ACTIVE PROTEINS (HYDROPHOBINS) OF MARINE FUNGI

Filamentous fungi synthesize and release a specific category of surface-active LMW proteins known as hydrophobins (Kwan et al. 2006; Linder et al. 2005; Sunde et al. 2008). In general, hydrophobins consist of 100–150 amino acids and share a conserved domain featuring eight cysteine residues (<20 kDa) (Kulkarni,

Nene and Joshi 2017; Wosten 2001). These hydrophobins serve diverse functions within fungi (Linder et al. 2005). Hydrophobins are assumed to be first released in a soluble state before spontaneously migrating to the hydrophilic–hydrophobic interface. At this interface, HBFs undergo self-assembly, forming amphipathic layers with diverse solubility (Cai et al. 2021; Hektor and Scholtmeijer 2005). These amphipathic layers help to reduce the interfacial tension, which permits the hyphae to breach the liquid surface and grow into the air by forming floating colonies. Moreover, aerial hyphae, spores, and fruiting bodies are also largely coated by hydrophobins to reduce wetting and provide resilience to environmental stresses (Cai et al. 2020; Cai et al. 2021).

In general, hydrophobins are categorized into two major groups based on their self-assembling behaviors as Class I and Class II (Wosten 2001). The Class I hydrophobins are characterized by cross β-structure and form extremely insoluble aggregates that resemble discrete rodlets, just like amyloid fibrils. While Class II hydrophobins produce less stable polymers that are soluble in certain organic solvents or SDS aqueous solutions, and lack the rodlet look of Class I hydrophobins, these assemblies exhibit exceptional stability and may be depolymerized in 100% trifluoroacetic acid (Chang, Choi and Na 2020; Lo et al. 2014). Moreover, Class I hydrophobins, further categorized into two subclasses as Class IA and Class IB, based on fungal species, exhibit diverse and lengthy amino acid sequences. In contrast, Class II hydrophobins typically feature short and uniform amino acid sequences (Chang, Choi and Na 2020; Gandier et al. 2017). Taken together, these hydrophobin protein film modifies the wetting characteristics of hydrophobic surfaces on liquids, gas bubbles, or solid materials, making them wettable, while simultaneously enabling the conversion of hydrophilic surfaces to hydrophobic (Wosten and Scholtmeijer 2015). These versatile properties of hydrophobins make them ideal elements for technical and medical applications (Cox and Hooley 2009; Scholtmeijer, Wessels and Wosten 2001). Notably, hydrophobins find utility in dispersing hydrophobic materials, stabilizing foam in food products, and immobilizing a range of substances, including peptides, enzymes, cells, antibodies, and inorganic molecules, onto surfaces (Wosten and Scholtmeijer 2015). Simultaneously, hydrophobins can be employed to prevent the binding of molecules. Additionally, hydrophobins exhibit therapeutic potential as immunomodulators and can be utilized in the production of recombinant proteins (Berger and Sallada 2019). However, the widespread adoption of hydrophobins in large-scale applications may pose challenges, such as production costs associated with recombinant proteins and/or the substantial quantities needed for such applications (Bayry et al. 2012).

Recently, a number of studies reported the applications of hydrophobins identified from marine fungi. Cicatiello et al. (2016) attempted to screen hydrophobins producing marine fungi with 23 fungal strains. According to the authors, among the tested marine fungi, six new hydrophobins were identified and among these proteins, four belong to Class I, while two are classified as Class II hydrophobins (Cicatiello et al. 2016). Moreover, according to the results, Class I hydrophobin isolates were found to consistently alter the wettability of a crystalline silicon chip, and Class II hydrophobins exhibited an impressive emulsification capacity under the tested conditions (Table 4.2). In another study, Cicatiello et al.

TABLE 4.2
Summary of Surface-Active Proteins (Hydrophobins) Identified from Different Geographical Locations

No.	Name of Fungi	Place Collected	Class	Reference
1	*Penicillium chrysogenum*	Elba Island in the	II	Cicatiello et al. (2016)
2	*Roussoellaceae* sp. 2	Mediterranean Sea	I	Cicatiello et al. (2016)
3	*Acremonium sclerotigenum*		I	Cicatiello et al. (2016, 2017)
4	*Myceliophthora verrucosa*		II	Cicatiello et al. (2016)
5	*Arthopyrenia salicis*		II	Cicatiello et al. (2016)
6	*Penicillium roseopurpureum*		I	Cicatiello et al. (2016, 2017)
7	*Aspergillus terreus*	Mediterranean Sea	I	Pitocchi et al. (2020)
8	*Trichoderma harzianum*	Mediterranean Sea	I	Pitocchi et al. (2020)
9	*Paradendryphiella salina*	Coquimbo Bay, Chile	II	Landeta-Salgado et al. (2021)
10	*Talaromyces pinophilus*	Montt in the South of Chile	I	Landeta-Salgado et al. (2021)

(2017) also reported the wettability properties and resistance to harsh washing of hydrophobins isolated from marine fungi called *Acremonium sclerotigenum* and *Penicillium roseopurpureum*. Pitocchi et al. (2020) reported the biosurfactant and emulsifier capabilities of two hydrophobins derived from two fungal strains, *A. terreus,* and *Trichoderma harzianum*, identified from the Mediterranean Sea (Pitocchi et al. 2020). According to the authors, *T. harzianum* hydrophobins were found to possess promising properties in reducing surface tension, via possessing both surfactant and emulsifying activities. Furthermore, Landeta-Salgado, Cicatiello, and Lienqueo (2021) described the use of mycoprotein and hydrophobin-like protein secreted by the marine fungus *Paradendryphiella salina* in submerged fermentation with the green seaweed *Ulva* spp. (Landeta-Salgado, Cicatiello and Lienqueo 2021). Additionally, Landeta-Salgado, Cicatiello, and Lienqueo (2021), also reported the production and characterization of cerato-platanin isolated from *Paradendryphiella salina* and two Class I hydrophobins from *Talaromyces pinophilus* (Landeta-Salgado et al. 2021). Taken together, these self-assembling proteins obtained from marine fungi have broad implications in different fields such as biotechnology, pharmacy, nanotechnology, and biomedicine (Figure 4.2).

4.4   FUTURE DIRECTIONS

Marine fungi have been a subject of study since the first recorded observation of a marine species. However, despite their widespread presence and abundance, marine fungi have received limited attention. The isolation of these fungi into pure cultures remains the established method for discovering and characterizing new taxa. The utilization of different methods for isolation, numerous culture media, different incubation temperatures, and the use of damp/moist chambers for incubation all can contribute to the enhancement of fungal diversity. Environmental factors, such

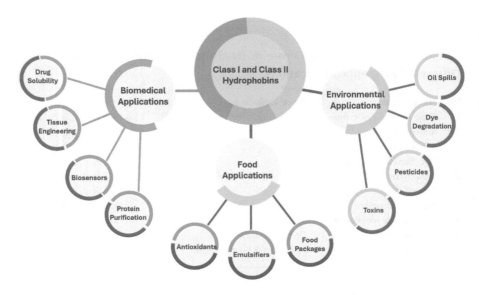

FIGURE 4.2   Potential applications of marine fungi-based hydrophobin.

as seasonal changes, can also lead to variations in fungal diversity. However, it is important to note that climate changes can exert a significant impact on the dynamics of marine fungal communities, potentially altering their structure and resulting in unforeseen consequences. Given the widely accepted global climate change, it is imperative to comprehensively understand how these changes affect the dynamics of marine fungi.

REFERENCES

Abdel-Lateff, A. 2008. "Chaetominedione, a new tyrosine kinase inhibitor isolated from the algicolous marine fungus *Chaetomium* sp." *Tetrahedron Letters* 49 (45):6398–6400. doi: 10.1016/j.tetlet.2008.08.064

Abdollahi, S., Z. Raoufi, I. Faghiri, A. Savari, Y. Nikpour, and A. Mansouri. 2013. "Contamination levels and spatial distributions of heavy metals and PAHs in surface sediment of Imam Khomeini Port, Persian Gulf, Iran." *Marine Pollution Bulletin* 71 (1–2):336–345. doi: 10.1016/j.marpolbul.2013.01.025

Agrawal, S., A. Adholeya, C. J. Barrow, and S. K. Deshmukh. 2018. "Marine fungi: An untapped bioresource for future cosmeceuticals." *Phytochemistry Letters* 23:15–20. doi: 10.1016/j.phytol.2017.11.003

Alao, M. B., and E. A. Adebayo. 2022. "Fungi as veritable tool in bioremediation of polycyclic aromatic hydrocarbons-polluted wastewater." *Journal of Basic Microbiology* 62 (3–4):223–244. doi: 10.1002/jobm.202100376

Alvarez-Barragan, J., C. Cravo-Laureau, L. Y. Wick, and R. Duran. 2021. "Fungi in PAH-contaminated marine sediments: Cultivable diversity and tolerance capacity towards PAH." *Marine Pollution Bulletin* 164:112082. doi: 10.1016/j.marpolbul.2021.112082

Ambade, B., S. S. Sethi, B. Giri, J. K. Biswas, and K. Bauddh. 2022a. "Characterization, behavior, and risk assessment of polycyclic aromatic hydrocarbons (PAHs) in the estuary sediments." *Bulletin of Environmental Contamination and Toxicology* 108 (2):243–252. doi: 10.1007/s00128-021-03393-3

Ambade, B., S. S. Sethi, S. Kurwadkar, P. Mishra, and L. Tripathee. 2022b. "Accumulation of polycyclic aromatic hydrocarbons (PAHs) in surface sediment residues of Mahanadi River estuary: Abundance, source, and risk assessment." *Marine Pollution Bulletin* 183:114073. doi: 10.1016/j.marpolbul.2022.114073

Amend, A., G. Burgaud, M. Cunliffe, V. P. Edgcomb, C. L. Ettinger, M. H. Gutiérrez, J. Heitman, E. F. Y. Hom, G. Ianiri, A. C. Jones, M. Kagami, K. T. Picard, C. A. Quandt, S. Raghukumar, M. Riquelme, J. Stajich, J. Vargas-Muñiz, A. K. Walker, O. Yarden, A. S. Gladfelter, and D. A. Garsin. 2019. "Fungi in the marine environment: Open questions and unsolved problems." *mBio* 10 (2). doi: 10.1128/mBio.01189-18

Babich, H., and G. Stotzky. 1983. "Nickel toxicity to estuarine/marine fungi and its amelioration by magnesium in sea water." *Water, Air, and Soil Pollution* 19 (2):193–202. doi: 10.1007/bf00211805

Baltar, F., Z. Zhao, and G. J. Herndl. 2021. "Potential and expression of carbohydrate utilization by marine fungi in the global ocean." *Microbiome* 9 (1):106. doi: 10.1186/s40168-021-01063-4

Bayry, J., V. Aimanianda, J. I. Guijarro, M. Sunde, and J. P Latge. 2012. "Hydrophobins–unique fungal proteins." *PLoS Pathogens* 8 (5):e1002700. doi: 10.1371/journal.ppat.1002700

Behera, A. D., and S. Das. 2023. "Ecological insights and potential application of marine filamentous fungi in environmental restoration." *Rev Environmental Science and Bio/Technology* 22 (2):281–318. doi: 10.1007/s11157-023-09655-2

Bengtson, S., B. Rasmussen, M. Ivarsson, J. Muhling, C. Broman, F. Marone, M. Stampanoni, and A. Bekker. 2017. "Fungus-like mycelial fossils in 2.4-billion-year-old vesicular basalt." *Nature Ecology and Evolution* 1 (6):141. doi: 10.1038/s41559-017-0141

Berbee, M. L., C. Strullu-Derrien, P. M. Delaux, P. K. Strother, P. Kenrick, M. A. Selosse, and J. W. Taylor. 2020. "Genomic and fossil windows into the secret lives of the most ancient fungi." *Nature Reviews Microbiology* 18 (12):717–730. doi: 10.1038/s41579-020-0426-8

Berger, B. W., and N. D. Sallada. 2019. "Hydrophobins: Multifunctional biosurfactants for interface engineering." *Journal of Biological Engineering* 13 (1):10. doi: 10.1186/s13036-018-0136-1

Birolli, W. G., A. Santos D. de, N. Alvarenga, Acfs Garcia, P. C. Romao, and M. Porto. 2018. "Biodegradation of anthracene and several PAHs by the marine-derived fungus *Cladosporium* sp. CBMAI 1237." *Marine Pollution Bulletin* 129 (2):525–533. doi: 10.1016/j.marpolbul.2017.10.023

Cai, F., R. Gao, Z. Zhao, M. Ding, S. Jiang, C. Yagtu, H. Zhu, J. Zhang, T. Ebner, M. Mayrhofer-Reinhartshuber, P. Kainz, K. Chenthamara, G. B. Akcapinar, Q. Shen, and I. S. Druzhinina. 2020. "Evolutionary compromises in fungal fitness: Hydrophobins can hinder the adverse dispersal of conidiospores and challenge their survival." *ISME Journal* 14 (10):2610–2624. doi: 10.1038/s41396-020-0709-0

Cai, F., Z. Zhao, R. Gao, P. Chen, M. Ding, S. Jiang, Z. Fu, P. Xu, K. Chenthamara, Q. Shen, G. Bayram Akcapinar, and I. S. Druzhinina. 2021. "The pleiotropic functions of intracellular hydrophobins in aerial hyphae and fungal spores." *PLoS Genetics* 17 (11):e1009924. doi: 10.1371/journal.pgen.1009924

Calado Mda, L., L. Carvalho, K. L. Pang, and M. Barata. 2015. "Diversity and ecological characterization of sporulating higher filamentous marine fungi associated with *Spartina maritima* (Curtis) Fernald in two Portuguese salt marshes." *Microbial Ecology* 70 (3):612–33. doi: 10.1007/s00248-015-0600-0

Chang, H. J., H. Choi, and S. Na. 2020. "Predicting the self-assembly film structure of Class II hydrophobin NC2 and estimating its structural characteristics." *Colloids and Surfaces B: Biointerfaces* 195:111269. doi: 10.1016/j.colsurfb.2020.111269

Chen, F., N. Tan, X. M. Yan, S. K. Yang, Z. G. She, and Y. C. Lin. 2014. "Uranium(VI) removal from aqueous solution by poly(amic acid)-modified marine fungus." *Separation Science and Technology* 49 (8):1251–1258. doi: 10.1080/01496395.2013.877033

Cicatiello, P., P. Dardano, M. Pirozzi, A. M. Gravagnuolo, L. De Stefano, and P. Giardina. 2017. "Self-assembly of two hydrophobins from marine fungi affected by interaction with surfaces." *Biotechnology and Bioengineering* 114 (10):2173–2186. doi: 10.1002/bit.26344

Cicatiello, P., A. M. Gravagnuolo, G. Gnavi, G. C. Varese, and P. Giardina. 2016. "Marine fungi as source of new hydrophobins." *International Journal of Biological Macromolecules* 92:1229–1233. doi: 10.1016/j.ijbiomac.2016.08.037

Cox, P. W., and P. Hooley. 2009. "Hydrophobins: New prospects for biotechnology." *Fungal Biology Reviews* 23 (1-2):40–47. doi: 10.1016/j.fbr.2009.09.001

da Silva Junior, F. C., M. Felipe, D. E. F. Castro, S. Araujo, H. C. N. Sisenando, and S. R. Batistuzzo de Medeiros. 2021. "A look beyond the priority: A systematic review of the genotoxic, mutagenic, and carcinogenic endpoints of non-priority PAHs." *Environmental Pollution* 278:116838. doi: 10.1016/j.envpol.2021.116838

Ekblad, A., H. Wallander, D. L. Godbold, C. Cruz, D. Johnson, P. Baldrian, R. G. Björk, D. Epron, B. Kieliszewska-Rokicka, R. Kjøller, H. Kraigher, E. Matzner, J. Neumann, and C. Plassard. 2013. "The production and turnover of extramatrical mycelium of ectomycorrhizal fungi in forest soils: Role in carbon cycling." *Plant and Soil* 366 (1–2):1–27. doi: 10.1007/s11104-013-1630-3

Freitas, F., C. Roca, and M. A. M. Reis. 2015. "Fungi as Sources of Polysaccharides for Pharmaceutical and Biomedical Applications." In *Handbook of Polymers for Pharmaceutical Technologies*, edited by V. K. Thakur and M. K. Thakur, 61–103. Wiley-Scrivener.

Gadd, G. M. 2007. "Geomycology: Biogeochemical transformations of rocks, minerals, metals and radionuclides by fungi, bioweathering and bioremediation." *Mycological Research* 111 (Pt 1):3–49. doi: 10.1016/j.mycres.2006.12.001

Gandier, J. A., D. N. Langelaan, A. Won, K. O'Donnell, J. L. Grondin, H. L. Spencer, P. Wong, E. Tillier, C. Yip, S. P. Smith, and E. R. Master. 2017. "Characterization of a Basidiomycota hydrophobin reveals the structural basis for a high-similarity class I subdivision." *Scientific Reports* 7 (1):45863. doi: 10.1038/srep45863

Gao, R., R. Liu, and C. Sun. 2022. "A marine fungus *Alternaria alternata* FB1 efficiently degrades polyethylene." *Journal of Hazardous Materials* 431:128617. doi: 10.1016/j.jhazmat.2022.128617

Hektor, H. J., and K. Scholtmeijer. 2005. "Hydrophobins: Proteins with potential." *Current Opinion in Biotechnology* 16 (4):434–9. doi: 10.1016/j.copbio.2005.05.004

Hyde, K. D., E. B. Gareth Jones, E. Leaño, S. B. Pointing, A. D. Poonyth, and L. L. P. Vrijmoed. 1998. "Role of fungi in marine ecosystems." *Biodiversity and Conservation* 7 (9):1147–1161. doi: 10.1023/a:1008823515157

Jennings, D. H. 1983. "Some aspects of the physiology and biochemistry of marine fungi." *Biological Reviews* 58 (3):423–459. doi: 10.1111/j.1469-185X.1983.tb00395.x

Jones, E. B. G., K.-L. Pang, M. A. Abdel-Wahab, B. Scholz, K. D. Hyde, T. Boekhout, R. Ebel, M. E. Rateb, L. Henderson, J. Sakayaroj, S. Suetrong, M. C. Dayarathne, V. Kumar, S. Raghukumar, K. R. Sridhar, A. H. A. Bahkali, F. H. Gleason, and C. Norphanphoun. 2019. "An online resource for marine fungi." *Fungal Diversity* 96 (1):347–433. doi: 10.1007/s13225-019-00426-5

Kohlmeyer, J, and E. Kohlmeyer. 1979. Marine Mycology, Elsevier Inc.

Kornprobst, J.-M. 2014. Encyclopedia of Marine Natural Products, Wiley-VCH Verlag GmbH.

Kulkarni, S., S. Nene, and K. Joshi. 2017. "Production of hydrophobins from fungi." *Process Biochemistry* 61:1–11. doi: 10.1016/j.procbio.2017.06.012

Kumar, A., V. Kumar, and J. Singh. 2019. "Role of Fungi in the Removal of Heavy Metals and Dyes from Wastewater by Biosorption Processes." In *Recent Advancement in White Biotechnology Through Fungi*, edited by A. N. Yadav, S. Singh, S. Mishra and A. Gupta, 397–418. Cham: Springer International Publishing.

Kwan, A. H., R. D. Winefield, M. Sunde, J. M. Matthews, R. G. Haverkamp, M. D. Templeton, and J. P. Mackay. 2006. "Structural basis for rodlet assembly in fungal hydrophobins." *Proceedings of the National Academy of Sciences* 103 (10):3621–3626. doi: 10.1073/pnas.0505704103

Landeta-Salgado, C., P. Cicatiello, and M. E. Lienqueo. 2021. "Mycoprotein and hydrophobin like protein produced from marine fungi *Paradendryphiella salina* in submerged fermentation with green seaweed *Ulva* spp." *Algal Research* 56:102314. doi: 10.1016/j.algal.2021.102314

Landeta-Salgado, C., P. Cicatiello, I. Stanzione, D. Medina, I. Berlanga Mora, C. Gomez, and M. E. Lienqueo. 2021. "The growth of marine fungi on seaweed polysaccharides produces cerato-platanin and hydrophobin self-assembling proteins." *Microbiological Research* 251:126835. doi: 10.1016/j.micres.2021.126835

Linder, M. B., G. R. Szilvay, T. Nakari-Setala, and M. E Penttila. 2005. "Hydrophobins: The protein-amphiphiles of filamentous fungi." *FEMS Microbiology Reviews* 29 (5): 877–896. doi: 10.1016/j.femsre.2005.01.004

Lo, V. C., Q. Ren, C. L. L. Pham, V. K. Morris, A. H. Kwan, and M. Sunde. 2014. "Fungal hydrophobin proteins produce self-assembling protein films with diverse structure and chemical stability." *Nanomaterials* 4 (3):827–843. doi: 10.3390/nano4030827

Mahajan, M., D. Manek, N. Vora, R. K. Kothari, C. Mootapally, and N. M. Nathani. 2021. "Fungi with high ability to crunch multiple polycyclic aromatic hydrocarbons (PAHs) from the pelagic sediments of Gulfs of Gujarat." *Marine Pollution Bulletin* 167:112293. doi: 10.1016/j.marpolbul.2021.112293

Marco-Urrea, E., I. Garcia-Romera, and E. Aranda. 2015. "Potential of non-ligninolytic fungi in bioremediation of chlorinated and polycyclic aromatic hydrocarbons." *New Biotechnology* 32 (6):620–628. doi: 10.1016/j.nbt.2015.01.005

Mushroom Research Foundation. 2024. "Marine fungi." https://www.marinefungi.org/ (accessed May 1, 2024).

Pang, K.-L., D. P. Overy, E. B. Gareth Jones, M. da Luz Calado, G, Burgaud, A. K. Walker, J. A. Johnson, R. G. Kerr, H.-J. Cha, and G. F. Bills. 2016. "'Marine fungi' and 'marine-derived fungi' in natural product chemistry research: Toward a new consensual definition." *Fungal Biology Reviews* 30 (4):163–175. doi: 10.1016/j.fbr.2016.08.001

Pitocchi, R., P. Cicatiello, L. Birolo, A. Piscitelli, E. Bovio, G. C. Varese, and P. Giardina. 2020. "Cerato-platanins from marine fungi as effective protein biosurfactants and bioemulsifiers." *International Journal of Molecular Sciences* 21 (8):2913. doi: 10.3390/ijms21082913

Raghukumar, S. 2017. "The Marine Environment and the Role of Fungi." In *Fungi in Coastal and Oceanic Marine Ecosystems*, 17–38. Cham: Springer International Publishing.

Rengarajan, T., P. Rajendran, N. Nandakumar, B. Lokeshkumar, P. Rajendran, and I. Nishigaki. 2015. "Exposure to polycyclic aromatic hydrocarbons with special focus on cancer." *Asian Pacific Journal of Tropical Biomedicine* 5 (3):182–189. doi: 10.1016/s2221-1691(15)30003-4

Richards, T. A., M. D. Jones, G. Leonard, and D. Bass. 2012. "Marine fungi: Their ecology and molecular diversity." *Annual Review of Marine Science* 4:495–522. doi: 10.1146/annurev-marine-120710-100802

Sanjeewa, K. K., I. P. Fernando, E. A. Kim, G. Ahn, Y. Jee, and Y. J. Jeon. 2017. "Anti-inflammatory activity of a sulfated polysaccharide isolated from an enzymatic digest of brown seaweed *Sargassum horneri* in RAW 264.7 cells." *Nutrition Research and Practice* 11 (1):3–10. doi: 10.4162/nrp.2017.11.1.3

Sanjeewa, K. K. A., K. H. I. N. M. Herath, Y.-S. Kim, Y.-J. Jeon, and S.-K. Kim. 2023. "Enzyme-assisted extraction of bioactive compounds from seaweeds and microalgae." *Trends in Analytical Chemistry* 167:117266. doi: 10.1016/j.trac.2023.117266

Scholtmeijer, K., J. G. Wessels, and H. A. Wosten. 2001. "Fungal hydrophobins in medical and technical applications." *Applied Microbiology and Biotechnology* 56 (1–2):1–8. doi: 10.1007/s002530100632

Sharif, N., A. Bibi, N. Zubair, and N. Munir. 2022. "Heavy Metal Accumulation Potential of Aquatic Fungi." In *Freshwater Mycology*, edited by S. A. Bandh and S. Shafi, 193–208. Elsevier.

Srivastava, S., and M. Kumar. 2019. "Biodegradation of Polycyclic Aromatic Hydrocarbons (PAHs): A Sustainable Approach." In *Sustainable Green Technologies for Environmental Management*, edited by S. Shah, V. Venkatramanan and R. Prasad, 111–139. Singapore: Springer.

Sunde, M., A. H. Kwan, M. D. Templeton, R. E. Beever, and J. P. Mackay. 2008. "Structural analysis of hydrophobins." *Micron* 39 (7):773–784. doi: 10.1016/j.micron.2007.08.003

Taboski, M. A., T. G. Rand, and A Piorko. 2005. "Lead and cadmium uptake in the marine fungi *Corollospora lacera* and *Monodictys pelagica*." *FEMS Microbiology Ecology* 53 (3):445–453. doi: 10.1016/j.femsec.2005.02.009

Tolosa, I., S. J. de Mora, S. W. Fowler, J. P. Villeneuve, J. Bartocci, and C. Cattini. 2005. "Aliphatic and aromatic hydrocarbons in marine biota and coastal sediments from the gulf and the Gulf of Oman." *Marine Pollution Bulletin* 50 (12):1619–1633. doi: 10.1016/j.marpolbul.2005.06.029

Vala, A. K. 2010. "Tolerance and removal of arsenic by a facultative marine fungus *Aspergillus candidus*." *Bioresource Technology* 101 (7):2565–2567. doi: 10.1016/j.biortech.2009.11.084

Wosten, H. A. 2001. "Hydrophobins: Multipurpose proteins." *Annual Review of Microbiology* 55 (1):625–646. doi: 10.1146/annurev.micro.55.1.625

Wosten, H. A., and K. Scholtmeijer. 2015. "Applications of hydrophobins: Current state and perspectives." *Applied Microbiology and Biotechnology* 99 (4):1587–1597. doi: 10.1007/s00253-014-6319-x

Yamini, V., and V. D. Rajeswari. 2023. "Metabolic capacity to alter polycyclic aromatic hydrocarbons and its microbe-mediated remediation." *Chemosphere* 329:138707. doi: 10.1016/j.chemosphere.2023.138707

Zain Ul Arifeen, M., Y. Ma, T. Wu, C. Chu, X. Liu, J. Jiang, D. Li, Y. R. Xue, and C. H Liu. 2022. "Anaerobic biodegradation of polycyclic aromatic hydrocarbons (PAHs) by fungi isolated from anaerobic coal-associated sediments at 2.5 km below the seafloor." *Chemosphere* 303 (Pt 2):135062. doi: 10.1016/j.chemosphere.2022.135062

5 Marine Microalgae

5.1 INTRODUCTION

Microalgae, which are ancient life-forms, serve as the foundation of aquatic food chains (Ashraf, Ahmad and Lu 2023). Moreover, about one-half of global oxygen production and photosynthesis is accomplished by marine microalgae (Matsunaga et al. 2005; Singh and Ahluwalia 2012). In general, microalgae are mostly microscopic unicellular, and all biological processes are strongly interconnected within the same space (Galasso et al. 2019; Venkatesan, Manivasagan and Kim 2015). Microalgae are characterized by their phylogenetic diversity, encompassing various phyla and classes of organisms, with some including cyanobacteria. Habitats of microalgae are not limited to aquatic ecosystems (freshwater, seawater, and hypersaline) but are also in moist soils and rocky substrates (Dittami et al. 2017; Vinokurova 2023).

As mentioned earlier, compared to terrestrial plants, the proliferation rate of microalgae is high, thus microalgae can be considered as a potential bioresource for renewable and highly quantitative production. The photosynthetic efficiency of microalgae varies between 10% and 20%, in contrast to 1–2% typically observed in most terrestrial plants. Notably, certain algal species, in the exponential growth phase, can double their biomass in periods as brief as 3.5 hours (Maeda et al. 2018; Singh and Ahluwalia 2012). To date, microalgae have demonstrated a range of industrial and commercial applications, such as pigments, food products, animal feed formulations, cosmetics, health supplements, nutraceuticals, agrochemicals, and fertilizers (Spolaore et al. 2006). In addition, a number of studies reported the potential of microalgae in wastewater treatment applications and in biofuel production applications (Peng et al. 2019). However, most of the research findings with commercial value are not being commercialized due to conditions like limited market size, high production costs, other microorganisms (fungi and bacteria), fossil resources, quality standards, safety, and environmental impact.

Other than the food source, the application of microalgae as an energy source has garnered attention for several compelling reasons (Norsker et al. 2011). Specifically, microalgae were found to produce a high yield from a given space compared to the land plants, and some strains such as *Nannochloropsis* spp. are exhibiting a high percentage of oil content (Ajjawi et al. 2017; Brennan and Regan 2020; Qiu et al. 2019). According to the previous reports in 2020, the global microalgae market reached an estimated value of US$3.4 billion and is expected to reach to US$4.6 billion by 2027, exhibiting a compound annual growth rate (CAGR) of 4.3% (Loke Show 2022). This expansion of microalgae-related market statistics demonstrates the growing recognition of the valuable functional ingredients inherent in microalgae and their sustainability advantages over the traditional methods in food and energy production (Silva et al. 2020). Other than the high lipid content and potential applications in food and other related industries, microalgae cultivation offers the advantage of low water consumption, and there exists the potential for production on arid lands (Janssen, Wijffels and Barbosa 2022). Thus, microalgae have been introduced as a raw material for many industries, such as biofertilizer processing, biochemicals, and biochar for wastewater treatment (Rahman and Melville 2023). This chapter aims to explore the potential of marine microalgae and their prospective applications in different industries such as nutraceuticals and functional foods, considering their sustainable nature and potential as food and pharmaceutical ingredients. In addition, general information such as cultivation and harvesting techniques and global production stats are also briefly discussed.

5.2   CULTIVATION OF MICROALGAE

In general, phototrophic microalgae absorb sunlight and assimilate CO_2 from the atmosphere and nutrients from the aquatic ecosystems. Therefore, artificial production systems should strive to imitate and enhance the optimal conditions of natural growth as much as possible (Brennan and Owende 2010). In general, microalgae cultivation systems are categorized into two systems: closed and open systems. Open systems use sunlight and are extensively exposed to the environment, providing a significant advantage by utilizing free natural resources. Closed systems, also known as photobioreactors (PBRs), come in various forms such as tubular, column, membrane, and flat plate PBRs (Xu et al. 2009). However, both open and closed systems have advantages and drawbacks.

5.2.1   OPEN SYSTEMS

Open ponds are the most common microalgae cultivation methods in commercial microalgal cultivation, although the applications of open ponds are restricted to specific types of microalgae adapted to extreme environmental conditions. This includes microalgae that thrive in environments characterized by high alkalinity: for example, *Spirulina*; those requiring high nutrient concentrations, as seen in *Chlorella*; and those well-suited to high-salinity conditions, like *Dunaliella*.

The commonly employed systems encompass circular ponds with rotating components for mixing, tanks, large shallow ponds, and raceway ponds. Open ponds are often designed with a similar configuration to raceway ponds (Figure 5.1). A raceway pond is constructed as a closed-loop recirculation channel, usually around 30 cm deep, where water circulation and mixing are facilitated by a paddle wheel (Qin, Alam and Wang 2019).

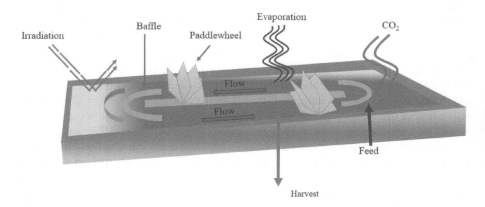

FIGURE 5.1   Detail structure of raceway.

TABLE 5.1
Comparison of Open and Closed Microalgae Cultivation Systems

	Open System	Closed System
Risk of contaminations	High	Low
CO_2 losses	High	Low
Percentages of evaporative losses	High	Low
Efficiency of light use	Week	High
Area/volume ratio	Low	High
Space utilization	Low	High
Scaling up possibility	High	Low
Process control	Difficult to manage	Easy to manage
Biomass production from a unit area	Low	High
Investment costs and return of investment	Low	High
Overall operational cost	Low	High
Harvesting cost	High	Relatively low

Despite being cost-effective to install and operate, open systems encounter various challenges, including non-axenic cultures (Narala et al. 2016). This non-axenic nature makes the cultures susceptible to contaminants that may outcompete the desired algal species. Additionally, predators like rotifers can pose a threat to the final yield, and unpredictable weather conditions can complicate the precise control of growth factors such as CO_2 levels, nutrients, and light intensity (Supriyanto et al. 2018). In addition, open pond systems want specific environmental conditions and control mechanisms to avoid contamination and pollution from other microalgae species and protozoans (Pulz and Scheibenbogen 1998; Wang et al. 2018). The advantages and disadvantages of open and closed systems are summarized in Table 5.1.

5.2.2 CLOSED SYSTEMS

Among the closed systems, photobioreactors offer a superior alternative to open cultivation systems for high-quality production due to their precisely controlled conditions (Singh and Sharma 2012). This allows for strain-specific optimization and overcomes the limitations of open systems. Photobioreactors play a crucial role in the cultivation of phototrophic microorganisms for the purpose of CO_2 fixation and the production of desired products (Chang et al. 2017). The design of PBRs hinges on two essential factors: the utilization of light for energy through penetration and distribution, and the provision of CO_2 as the inorganic carbon source. These two elements stand as fundamental considerations in shaping the design of effective photobioreactors (Chang et al. 2017).

In general, photobioreactors provide space efficiency, maximize light availability, and drastically reduce contamination risks. However, issues such as biofouling, overheating, benthic algae growth, cleaning difficulties, and oxygen buildup lead to growth limitations, and most significantly, high capital costs for design and operation are major drawbacks of CPBRs. Vertical tubular, bubble column, airlift, flat panel, horizontal tubular, helical, and stirred tank are some examples for photo-bioreactors (Singh and Sharma 2012; Yen et al. 2019).

5.3 MICROALGAE HARVESTING TECHNIQUES

Microalgae biomass, recognized as a potential bioenergy feedstock, is believed to possess significant potential for biofuel production. Traditionally, the harvesting of microalgae biomass involves a diverse array of physical, chemical, electrical, magnetic, and biological processes (Ananthi et al. 2021; Roy and Mohanty 2019). These methods can be applied independently or in combination to enhance the efficiency of the harvesting process. In general, the major concern in the large-scale production of microalgae is identifying cost-effective harvesting and dewatering processes (typically a harvesting method which yields algal biomass with minimum moisture content is preferred) (Molina Grima et al. 2003). The commercialization of products developed from microalgae requires the implementation of harvesting techniques of microalgae in a manner that is both efficient and sustainable, while also being cost-effective to reduce the production cost of the final product. Physical microalgae harvesting techniques include filtration, flotation, gravity sedimentation, centrifugation, electrocoagulation, and magnetic flocculation processes (Figure 5.2). These techniques exhibit high recovery rates for microalgae harvesting and most of these novel harvesting techniques typically yield contamination-free biomass eventually suitable for use as raw material for high-value products. However, they are plagued by drawbacks such as elevated operational costs, high-energy requirements, and extended durations for harvesting. The substantial cost associated with harvesting, typically accounting for 20–30% of overall production costs, poses a significant hindrance to the progress of microalgae

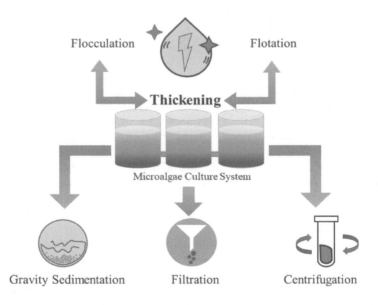

FIGURE 5.2   Microalgae harvesting techniques.

biotechnology industries (Esteves et al. 2020; Xu et al. 2021). In the following sections, major microalgae harvesting methods are briefly described.

5.3.1 HARVESTING TECHNIQUES OF MICROALGAE

Due to the low cell density, in most microalgae, a cost-effective and efficient recovery technique is required to identify and apply during the harvesting (Okoro et al. 2019). Previously, a number of studies highlighted the potential loss of biomass and lipid content due to the delays in harvesting of microalgae (Napan et al. 2015). Harvesting methods such as filtration, centrifugation, flotation, flocculation, gravity sedimentation, electrolytic processes, electrophoresis, and magnetic separation, are popular microalgae harvesting, each with its own set of advantages and disadvantages (Table 5.2) (Wang et al. 2015). It is important to note that a singular method may not achieve high efficiency for all microalgae species.

TABLE 5.2

Advantages and Disadvantages of Different Microalgae Harvesting Methods

Harvesting Method	Advantages	Disadvantages
Filtration	Availability of a range of filters and membranes, reliability, ease of application, minimal energy need, cell intracellular materials stay intact, and biomass recovery of 70–90%	Species-dependent, suitable for big species, fouling probable, high operating costs
Flotation	Good for large scale, may be integrated with the gaseous transfer, the process is faster than sedimentation, biomass recovery of 50–90%, and low cost and space required	Species-dependent, flocculants are needed
Centrifugation	Reliable and rapid method, possibility to recover more than 90% of biomass, applicable to all microalgae species, ideal for high-value product recovery	The high initial investment cost for equipment and maintenance, energy demand, risk of cell damage, unsuitable for large-scale
Flocculation	Comparatively a low-cost and flexible method, a wide range of flocculants available, possibility to recover more than 90% of the biomass, cell damage is minimal, energy consumption is low, and simple procedure	Chemical contamination, flocculant dosage may be high, removal of flocculants may be difficult, flocs may take longer time to settle, pH-dependent, reuse of culture medium is challenging
Electroflocculation	Low-energy requirement compared to the other methods such as centrifugation, biomass recovery of over 90%, can be applied to most microalgae species, no chemicals required	Metallic ion contamination, electrode replacement, chlorine can be formed by the treatment of seawater, high maintenance cost
Magnetic separation	The process is comparatively fast, low cost, possibility to combine with other methods, possibility to recover more than 90% of the biomass	Biomass requires modification, the separation of biomass is complex compared to the other methods

5.3.1.1   Physical Methods

Centrifugation, filtration, sedimentation, and flotation are some common examples of physical harvesting methods for microalgae. Each method relies on distinct principles such as centrifugal force, filtration through pores, gravity settling, and flotation with air bubbles. While efficient, each method has limitations, underscoring the need for careful consideration in the application (refer to Table 5.2).

5.3.1.1.1   Centrifugation

The centrifugation technique is employed for the separation of microalgae cells based on the density and particle size of individual components within the growth media. Despite its high efficiency in cell harvesting, significant drawbacks include prolonged processing time and elevated energy consumption (Deepa, Sowndhararajan and Kim 2023). However, the application of centrifugation carries the risk of causing cellular damage due to the exertion of high gravitational forces.

5.3.1.1.2   Filtration

Filtration is a physical process that can be used to separate solids from liquids or gases by introducing a medium that allows only the fluid to pass through. Microfilters, also known as strainers, with specific pore sizes are utilized in the filtration process (Enamala et al. 2018). These filters function as rotating filters and undergo frequent backwashing. In general, the filtration technique can be used to filter larger-sized algae, especially filamentous species (Deepa, Sowndhararajan and Kim 2023). The major disadvantage of filtration-aided microalgae harvesting is the frequent change of filters and membranes which makes the process economically not viable for commercial applications or value-addition applications (Enamala et al. 2018)

5.3.1.1.3   Sedimentation

Sedimentation emerges as a cost-effective and technically feasible approach for large-scale operations, serving the purpose of pre-concentrating the microalgal suspension before the dewatering/drying process (Leite and Daniel 2020). In commercial/large-scale systems, centrifugation is the most common harvesting method; the sedimentation followed by centrifugation has the potential to substantially decrease overall costs. Sedimentation and centrifugation would be employed as the harvesting and dewatering methods, respectively (Salim et al. 2011).

5.3.1.1.4   Flotation

Flotation is another microalgae separation process that involves the attachment of air or gas bubbles to solid particles. These bubbles adhere to and transport the solid particles toward the surface for collection (Gerardo et al. 2015). Successful flotation may be related to the suitable range of particle size between bubbles and particles as well as the probability of bubbles to particle attachments through collision or adhesion. Usually, coagulants or surfactants are introduced into the culture tanks to make the microalgal cells hydrophobic and expand the mass transfer between the air and the microalgal particles for the enhancement of particle separation (Japar, Takriff and Yasin 2017). The commonly utilized surfactants include aluminum sulfate $(Al_2(SO_4)_3)$, cetyltrimethylammonium bromide (CTAB), iron(III) sulfate $(Fe_2(SO_4)_3)$, chitosan, and iron(III) chloride $(FeCl_3)$ (Japar, Takriff and Yasin 2017).

5.3.1.2 Chemical Methods

Chemical microalgae harvesting methods include auto-flocculation and coagulation. In general, chemical agents or naturally occurring biopolymers are commonly used to induce microalgae aggregation. These methods provide alternatives that mitigate the energy-intensive nature of some physical techniques. However, challenges like chemical contamination and flocculant removal must be addressed (Difusa, Mohanty and Goud 2015).

5.3.1.2.1 Auto-Flocculation

When the pH of the medium is adjusted either upward or downward to a specific threshold, the cells aggregate and precipitate due to gravitational forces. Introducing additional bases or acids into the medium enhances the formation of compact flocs, leading to reduced settling times. It is essential to note that not all microalgae species exhibit flocculation in response to variations in pH levels (Augustine et al. 2019; Pérez et al. 2017).

5.3.1.2.2 Coagulation

Coagulation is another popular method employed in the harvesting of microalgae, involving the introduction of coagulants $(Al_2(SO_4)_3$, Synth), ferric chloride $(FeCl_3$, Qhemis), Tanfloc SG, and Zetag 8185) to facilitate the aggregation of microalgal cells. This process forms insoluble precipitates, aiding in the efficient separation of microalgae from the culture medium. Coagulation is a chemical method that enhances the sedimentation of microalgae, streamlining the harvesting process (Leite, Hoffmann and Daniel 2019).

5.3.1.3 Biological Methods

5.3.1.3.1 Bioflocculation

Bioflocculation employs nonchemical self-aggregating microorganisms (or their extracellular biopolymers) to induce flocculation and facilitate the harvest of non-flocculating target microalgae (Ummalyma et al. 2017). Bioflocculants encompass bacteria, fungi, yeasts, self-flocculating algae, and their exudate-rich culture supernatants. This technique's absence of chemical additives mirrors auto-flocculation, rendering it a sustainable and environment-friendly approach for microalgae harvesting (Matter et al. 2019).

5.3.1.3.2 Predation

This method utilizes zooplankton or other organisms to graze on microalgae populations, effectively controlling population size but potentially unsuitable for all applications.

5.3.1.3.3   Microbial Lysis

This technique employs bacteria or viruses to lyse microalgae cells and release the intracellular contents. While useful for extracting specific bioproducts, it may not be suitable for whole-cell harvesting.

5.4   APPLICATIONS OF MARINE MICROALGAE

The versatile nature of macroalgae extends beyond traditional uses and has found applications in various industries. Beyond their role as a dietary staple, macroalgae serve as valuable resources for animal feed, contributing to the agriculture sector (Salim et al. 2011). Additionally, their therapeutic properties have positioned them in the realm of medicine, further expanding their impact. The increasing global recognition of the benefits associated with macroalgae has spurred a surge in industrial applications, with ongoing research and development exploring new avenues for their utilization (Khalid 2020). These widespread adoption underscores the multifaceted potential of macroalgae across different sectors, contributing to their significance on a global scale. In this section, major commercial applications of microalgae are briefly discussed (Figure 5.3).

5.4.1   APPLICATIONS OF MICROALGAE IN THE FOOD INDUSTRY

The rising global population and climate changes will reduce the capacity of traditional food production systems to feed rising global populations in the future, potentially leading to a food shortage within the next few decades (Gu, Andreev, and Dupre 2021; van Dijk et al. 2021). Simultaneously, increasing environmental concerns necessitate a shift in conventional food supply systems. Among the possible alternatives to address these challenges, microalgae have been identified as promising candidates to fill the food supply and demand gap in the coming decades (Chen et al. 2022). In addition, with robust carbon sequestration capabilities,

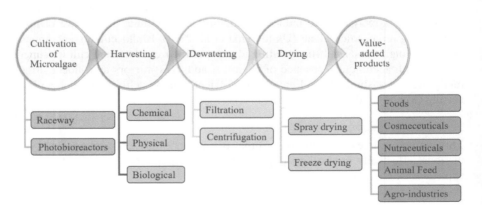

FIGURE 5.3   Major steps involved in microalgae processing and applications.

microalgae are poised to play a principal role in global carbon neutrality initiatives in the future (Prasad et al. 2021).

According to previous studies, similar to the land plants, microalgae are also rich in essential nutrients. Specifically, microalgal species such as *Chlorella* sp., *Spirulina plantesis, Dunaliela saline, Dunaliella terticola,* and *Aphanizomenon flos-aquae* are rich in essential amino acids and proteins (Liang et al. 2020). These microscopic organisms are also rich in vitamins like A, B (1, 2, 6, and 12), C, and E, along with essential minerals (Barta, Coman and Vodnar 2021) such as iron, calcium, potassium, magnesium, and iodine (Becker 2013). In addition, according to earlier studies, microalgae are promising candidates to extract omega-3 (ω-3) long-chain polyunsaturated fatty acids (LC-PUFAs), namely, eicosapentaenoic acid (EPA) and docosahexaenoic acid (DHA) (Tocher et al. 2019).

At present, food items developed from microalgae are promoted as health-promoting foods or functional products and are found in the market in various forms such as capsules, tablets, powders, and liquids. Furthermore, they are blended into a diverse range of products, including candies, gums, snacks, pastes, noodles, breakfast cereals, wine, and various other beverages However, at present, *Chlorella* and *Spirulina* species dominate the global microalgae market as they are gaining popularity in the health-food supermarkets and stores. This is attributed to the nutrient-rich profiles of these species. Moreover, fostering public awareness and understanding of the nutritional and environmental benefits of microalgae-based foods will play a crucial role in shaping consumer perceptions and preferences.

5.4.2 AGRICULTURAL APPLICATIONS

The applications of macroalgae in agriculture as fertilizers is one of the oldest traditions. Historically, people who lived in coastal areas all around the world were found to use microalgae as plant fertilizers in their agricultural fields (Griffiths et al. 2016). Microalgae extracts represent many applications in modern sustainable agriculture, such as enhancing nutrient uptake, improving crop productivity, preventing nutrient loss, enhancing physiological status, and addressing abiotic stress (Kapoore, Wood and Llewellyn 2021). In addition, previous studies have been conducted to assess the impact of microalgae extracts as biostimulants and biofertilizers under various cultivation conditions, including open fields, greenhouses, and hydroponics (Kumar and Singh 2020). These studies have been performed on a variety of crops, such as cereals, vegetables, and medicinal plants, revealing positive impacts (Dineshkumar et al. 2018; Plaza et al. 2018).

Moreover, microalgae extracts found to sustain agricultural productivity while minimizing environmental degradation (Alvarez et al. 2021). Besides, microalgae are also recognized for their potential as biocontrol agents due to their ability to exhibit antagonistic effects against various plant pathogens, including bacteria, fungi, and nematodes (Righini and Roberti 2019). This is primarily attributed to the production of hydrolytic enzymes and biocidal compounds such as benzoic acid, majusculonic acid, etc. (La Bella et al. 2022). These antimicrobial compounds exert their action by suppressing pathogenic microbes, disrupting the cytoplasmic membrane, and inhibiting protein synthesis, among other mechanisms (Renuka et al. 2018).

Additionally, microalgae have the potential to increase the production of agricultural crops because they contribute elements such as magnesium, strontium, boron, and iron. Moreover, microalgae extracts contain several natural plant hormones such as auxins, cytokinins, gibberellin, and others that can further improve plant growth while suppressing pathogens or acting as fertilizers (Righini and Roberti 2019). Taken together, microalgae have a huge role in modern sustainable agriculture.

5.4.3 PHARMACEUTICALS AND COSMETICS APPLICATIONS OF MICROALGAE

With properties such as autotrophic nature, fast growth, short reproduction cycle, and strong adaptability to environmental changes, microalgae become potential candidates for extracting bioactive compounds at a low cost for industries like cosmeceuticals and nutraceuticals (Zhuang et al. 2022). In response to emerging trends and the growing demand for natural bioactive compounds, microalgae, along with the bioactive ingredients they yield, are increasingly recognized as promising alternatives to traditional raw materials in the pharmaceutical and cosmetics industries (Balasubramaniam et al. 2021).

The diverse array of bioactive compounds found in microalgae, including carotenoids (astaxanthin, canthaxanthin, β-carotene, zeaxanthin, violaxanthin), polyunsaturated fatty acids (PUFA), phycobilin, single-cell proteins, amino acids, polysaccharides, vitamins, sterols, and enzymes, has gained considerable interest mainly in the nutraceutical and cosmeceutical industry (Andrade et al. 2021; Kim et al. 2021). These valuable bioactive metabolites hold immense potential for advancing the pharmaceutical and cosmetic industries (Zhuang et al. 2022).

5.4.3.1 Pharmaceutical Applications of Microalgae

Phycobilins (not found in higher plants) are light-harvesting pigments found in microalgae and cyanobacteria (Frank and Cogdell 2012). According to earlier studies, phycobilins (such as phycocyanin and phycoerythrin) isolated from microalgae were found to possess strong antioxidant, anti-inflammatory, and antidiabetic properties (Ramos-Romero et al. 2021). Moreover, PUFAs derived from microalgae, such as omega-3 fatty acids, have gained attention for their cardiovascular and neurological benefits. Microalgae, particularly species like *Spirulina* and *Chlorella*, are rich sources of PUFAs like DHA and EPA (Ramesh Kumar et al. 2019). These fatty acids are essential for brain health, and their pharmaceutical applications include neuroprotective interventions, antioxidant, anti-inflammatory, anti-obesity, antidiabetic, and cardiovascular health support (Remize et al. 2021).

5.4.3.2 Cosmetics Applications of Microalgae

Different types of bioactive compounds isolated from microalgae such as pigments (astaxanthin, beta-carotene, and carotenoids), lipids, PUFA, and polyphenols play promising roles in the cosmetic industry (Coulombier, Jauffrais and Lebouvier 2021). In cosmeceuticals, these bioactive compounds play a vital role in neutralizing free radicals, helping to protect the skin from oxidative stress caused by environmental factors like UV radiation and pollution (Gauthier, Senhorinho and Scott 2020). In addition, antioxidant properties in these compounds contribute to developing antiaging, skin-lightening, and depigmenting formulations.

Microalgae also contribute to cosmeceutical formulations through their omega-3 fatty acids. These essential fatty acids, including DHA and EPA, play a crucial role in maintaining skin health (Ibrahim et al. 2023; Kumari, Pabbi and Tyagi 2023). Omega-3 fatty acids contribute to the lipid barrier of the skin, promoting hydration and preventing moisture loss (Michalak et al. 2021). Formulations enriched with omega-3 fatty acids are sought after for their potential to enhance skin texture, reduce inflammation, and improve overall skin appearance. Furthermore, microalgae-derived proteins offer benefits in cosmetic applications (Lucakova, Branyikova and Hayes 2022). Proteins like collagen and elastin are vital for maintaining skin elasticity and firmness (Ruiz Martinez et al. 2020). While natural collagen and elastin proteins are often used in antiaging formulations, microalgae proteins can act as bioactive agents, supporting the skin's natural structure and contributing to a more youthful complexion.

5.4.4 Wastewater Treatment Applications of Microalgae

Microalgae present a promising solution for transforming slurry and wastewater nutrients into valuable resources as they can grow in marginal and contaminated wastewater (Janssen, Wijffels and Barbosa 2022). With modern sustainable development objectives, microalgae have the potential to play a vital role in wastewater treatment. Microalgae offers a sustainable solution to nutrient removal, organic pollutant degradation, and heavy metal sequestration. Specifically, microalgae are increasingly being identified as potential solutions to remove metal ion contaminants from wastewater through biosorption (Leong and Chang 2020). Removing heavy metal ions (lead (Pb), arsenic (As), chromium (Cr), cadmium (Cd), and mercury (Hg)) through bioabsorption offers positive advantages over the traditional metal removal techniques such as inexpensive and simple, eco-friendly, and can be performed over a different of experimental settings (Spain, Plohn and Funk 2021; Touliabah et al. 2022).

Besides heavy metal bioabsorption, the ability of microalgae to assimilate nutrients, including nitrogen and phosphorus, aids in preventing eutrophication. Microalgae contribute to bioremediation by absorbing and metabolizing various organic and inorganic pollutants such as nitrogen, phosphate, organic carbons, and pharmaceutical compounds, providing an eco-friendly approach to wastewater remediation (Zainith et al. 2021). According to earlier studies, microalgae genera, such as *Chlamydomonas*, *Nitzschia*, *Botryococcus*, *Chlorella*, *Scenedesmus*, and *Micractinium*, are found to be effective in waste treatment applications (Touliabah et al. 2022).

Additionally, certain microalgae species act as natural biosorbents and actively remove hydrocarbons (*Nannochloropsis oculata*), and dye compounds from industrial effluents (Marques et al. 2021). Moreover, their role in oxygenation, aeration, and the mitigation of carbon dioxide emissions adds further value to wastewater treatment processes. Integration with wastewater treatment plants, the implementation of Algal Turf Scrubber systems, and wastewater polishing (Chlorophyta phylum) underscore the versatility of microalgae in enhancing water quality (Couto et al. 2022). Despite challenges in scaling up processes and optimizing economic viability, ongoing research aims to identify robust microalgae strains and optimize cultivation conditions for efficient nutrient removal and pollutant uptake. Microalgae-based wastewater treatment holds promise as an environment-friendly solution, contributing to cleaner water resources and sustainable practices.

5.4.5   BIOFUEL PRODUCTION

The over and unsustainable use of fossil fuels has been strongly linked with the serious issues that we are facing at the moment, including global greenhouse effects, global warming, and environmental pollution. Moreover, the issues associated with fossil fuel burning have affected most macro- and microorganisms worldwide at an alarming rate (Alishah Aratboni et al. 2019).

Microalgal biomass has been demonstrated as a promising source to extract various biofuels such as biodiesel, bio-oil, biomethane, bioethanol, and more. An additional advantage of incorporating microalgal technologies into industrial processes is the capacity of microalgae to sequester CO_2 during application and biomass production, leading to a subsequent reduction in CO_2 emissions (Siddiki et al. 2022). Under ideal environmental conditions, the oil percentage of microalgae generally ranges between 10% and 30% from their dry weights. For example, microalgal species such as *Nannochloris* sp. (56%) and *Schizochytrium* sp. (80%) are reported to produce higher lipid content under optimum growth conditions (Sanjeewa et al. 2016; Shokravi et al. 2020). However, the growth rate of high lipid-containing microalgal species is often slow. Alternatively, microalgae species such as *Chlorella* sp. and *Scenedesmus* sp. with a higher growth rate were found to produce minor amounts of lipid content from their dry weights (Shokravi et al. 2020). Thus, taken together, the cultivation of microalgae for the biofuel industry requires the careful selection of strains known for their high lipid content and rapid growth rates. In general, different cultivation systems, such as open ponds, raceways, photobioreactors, and hybrid setups, can be used. However, highly efficient systems like photobioreactors are more suitable for commercial biofuel production systems (Suparmaniam et al. 2019).

Once the microalgae reach a sufficient biomass, they are harvested and the lipids are extracted. Several extraction techniques are available for lipid extraction from microalgae such as organic solvents-assisted extraction methods, supercritical fluids extraction, ionic liquids, deep eutectic solvents, and switchable solvents (Lee et al. 2020; Vasistha et al. 2021). The extracted lipids are then subjected to transesterification, a process where they react with alcohol and a catalyst to produce biodiesel (fatty acid methyl or ethyl esters) and glycerol (Menegazzo and Fonseca 2019).

In addition to biofuels, microalgae cultivation yields valuable co-products such as animal feed, fertilizers, and high-value chemicals, enhancing the economic viability of the overall process (Olguín et al. 2022). One of the notable advantages of microalgae-based biofuels is their high productivity, owing to the organisms' efficient biomass production. Furthermore, the carbon dioxide emitted during the combustion of biofuels is offset by the CO_2 absorbed during microalgae photosynthesis, potentially making the process carbon-neutral. Additionally, microalgae cultivation can occur on nonarable land or even in wastewater, minimizing competition with food crops for valuable agricultural space (Nwoba et al. 2020).

However, the industry faces certain challenges. Achieving economic viability necessitates the development of cost-effective large-scale cultivation and harvesting methods. Technological hurdles persist in optimizing processes such as lipid extraction, and scale-up issues must be addressed to ensure consistency and prevent contamination in the transition from lab-scale to commercial-scale production.

5.5 FUTURE DIRECTIONS

The potential of microalgae to reduce CO_2 pollution, their applications in wastewater treatment, and the production of biofuels hold promise for a more sustainable future. However, the industrial applications of microalgae face challenges, particularly in terms of energy input and production costs, which need resolution for commercial viability. To make microalgal biofuel production both economically and technologically feasible, it is crucial to also generate other essential bioproducts. Consideration must be given to reducing large quantities of biomass required for the extraction of products from microalgae in future research. Additionally, future microalgae-based research studies should be designed to achieve sustainable development goals and efforts should emphasize simplicity in processes applicable to industries such as pharmaceuticals, cosmetics, or biofuels, particularly in developing countries.

REFERENCES

Ajjawi, I., J. Verruto, M. Aqui, L. B. Soriaga, J. Coppersmith, K. Kwok, L. Peach, E. Orchard, R. Kalb, W. Xu, T. J. Carlson, K. Francis, K. Konigsfeld, J. Bartalis, A. Schultz, W. Lambert, A. S. Schwartz, R. Brown, and E. R. Moellering. 2017. "Lipid production in *Nannochloropsis gaditana* is doubled by decreasing expression of a single transcriptional regulator." *Nature Biotechnology* 35 (7):647–652. doi: 10.1038/nbt.3865

Alishah Aratboni, H., N. Rafiei, R. Garcia-Granados, A. Alemzadeh, and J. R. Morones-Ramirez. 2019. "Biomass and lipid induction strategies in microalgae for biofuel production and other applications." *Microbial Cell Factories* 18 (1):178. doi: 10.1186/s12934-019-1228-4

Alvarez, A. L., S. L. Weyers, H. M. Goemann, B. M. Peyton, and R. D. Gardner. 2021. "Microalgae, soil and plants: A critical review of microalgae as renewable resources for agriculture." *Algal Research* 54:102200. doi: 10.1016/j.algal.2021.102200

Ananthi, V., P. Balaji, R. Sindhu, S. H. Kim, A. Pugazhendhi, and A. Arun. 2021. "A critical review on different harvesting techniques for algal based biodiesel production." *Science of the Total Environment* 780:146467. doi: 10.1016/j.scitotenv.2021.146467

Andrade, D. S., H. F. Amaral, F. Z. Gavilanes, L. R. I. Morioka, J. M. Nassar, J. M. de Melo, H. R. Silva, and T. S. Telles. 2021. "Microalgae: Cultivation, Biotechnological, Environmental, and Agricultural Applications." In *Advances in the Domain of Environmental Biotechnology*, edited by N. R. Maddela, L. C. G. Cruzatty and S. Chakraborty, 635–701. Singapore: Springer.

Ashraf, N., F. Ahmad, and Y. Lu. 2023. "Synergy between microalgae and microbiome in polluted waters." *Trends in Microbiology* 31 (1):9–21. doi: 10.1016/j.tim.2022.06.004

Augustine, A., A. Tanwar, R. Tremblay, and S. Kumar. 2019. "Flocculation processes optimization for reuse of culture medium without pH neutralization." *Algal Research* 39:101437. doi: 10.1016/j.algal.2019.101437

Balasubramaniam, V., R. D. Gunasegavan, S. Mustar, J. C. Lee, and M. F. Mohd Noh. 2021. "Isolation of industrial important bioactive compounds from microalgae." *Molecules* 26 (4):943. doi: 10.3390/molecules26040943

Barta, D. G., V. Coman, and D. C. Vodnar. 2021. "Microalgae as sources of omega-3 polyunsaturated fatty acids: Biotechnological aspects." *Algal Research* 58:102410. doi: 10.1016/j.algal.2021.102410.

Becker, E. W. 2013. "Microalgae for Human and Animal Nutrition." In *Handbook of Microalgal Culture*, edited by A. Richmond, 461–503. Blackwell Publishing Ltd.

Brennan, L., and P. Owende. 2010. "Biofuels from microalgae: A review of technologies for production, processing, and extractions of biofuels and co-products." *Renewable and Sustainable Energy Reviews* 14 (2):557–577. doi: 10.1016/j.rser.2009.10.009

Brennan, B., and F. Regan. 2020. "In-situ lipid and fatty acid extraction methods to recover viable products from *Nannochloropsis* sp." *Science of the Total Environment* 748:142464. doi: 10.1016/j.scitotenv.2020.142464

Chang, J. S., P. L. Show, T. C. Ling, C. Y. Chen, S. H. Ho, C. H. Tan, D. Nagarajan, and W. N. Phong. 2017. "Photobioreactors." In *Current Developments in Biotechnology and Bioengineering*, edited by C. Larroche, M. Á. Sanromán, G. Du and A. Pandey, 313–352. Elsevier.

Chen, C., T. Tang, Q. Shi, Z. Zhou, and J. Fan. 2022. "The potential and challenge of micro-algae as promising future food sources." *Trends in Food Science & Technology* 126: 99–112. doi: 10.1016/j.tifs.2022.06.016

Coulombier, N., T. Jauffrais, and N. Lebouvier. 2021. "Antioxidant compounds from micro-algae: A review." *Marine Drugs* 19 (10). doi: 10.3390/md19100549

Couto, A. T., M. Cardador, S. Santorio, L. Arregui, B. Sicuro, A. Mosquera-Corral, P. M. L. Castro, and C. L. Amorim. 2022. "Cultivable microalgae diversity from a freshwater aquaculture filtering system and its potential for polishing aquaculture-derived water streams." *Journal of Applied Microbiology* 132 (2):1543–1556. doi: 10.1111/jam.15300

Deepa, P., K. Sowndhararajan, and S. Kim. 2023. "A review of the harvesting techniques of microalgae." *Water* 15 (17). doi: 10.3390/w15173074

de Souza Leite, L., M. T. Hoffmann, and L. A. Daniel. 2019. "Coagulation and dissolved air flotation as a harvesting method for microalgae cultivated in wastewater." *Journal of Water Process Engineering* 32:100947. doi: 10.1016/j.jwpe.2019.100947

Difusa, A., K. Mohanty, and V. V. Goud. 2015. "Advancement and Challenges in Harvesting Techniques for Recovery of Microalgae Biomass." In *Environmental Sustainability*, edited by P. Thangavel and G. Sridevi, 159–169. New Delhi, India: Springer.

Dineshkumar, R., J. Subramanian, A. Arumugam, A. Ahamed Rasheeq, and P. Sampathkumar. 2018. "Exploring the microalgae biofertilizer effect on onion cultivation by field experiment." *Waste and Biomass Valorization* 11 (1):77–87. doi: 10.1007/s12649-018-0466-8

Dittami, S. M., S. Heesch, J. L. Olsen, and J Collen. 2017. "Transitions between marine and freshwater environments provide new clues about the origins of multicellular plants and algae." *Journal of Phycology* 53 (4):731–745. doi: 10.1111/jpy.12547

Enamala, M. K., S. Enamala, M. Chavali, J. Donepudi, R. Yadavalli, B. Kolapalli, T. V. Aradhyula, J. Velpuri, and C. Kuppam. 2018. "Production of biofuels from microal-gae: A review on cultivation, harvesting, lipid extraction, and numerous applications of microalgae." *Renewable and Sustainable Energy Reviews* 94:49–68. doi: 10.1016/j.rser.2018.05.012

Esteves, A. F., C. J. Almeida, A. L. Gonçalves, and J. C. Pires. 2020. "Microalgae Harvesting Techniques." In *Handbook of Microalgae-Based Processes and Products*, edited by E. Jacob-Lopes, M. M. Maroneze, M. I. Queiroz, and L. Q. Zepka, 225–281. Academic Press.

Frank, H. A., and R. J. Cogdell. 2012. "8.6 Light Capture in Photosynthesis." In *Comprehensive Biophysics*, edited by E. H. Egelman, 94–114. Amsterdam: Elsevier.

Galasso, C., A. Gentile, I. Orefice, A. Ianora, A. Bruno, D. M. Noonan, C. Sansone, A. Albini, and C. Brunet. 2019. "Microalgal derivatives as potential nutraceutical and food sup-plements for human health: A focus on cancer prevention and interception." *Nutrients* 11 (6). doi: 10.3390/nu11061226

Gauthier, M. R., G. N. A. Senhorinho, and J. A. Scott. 2020. "Microalgae under environmental stress as a source of antioxidants." *Algal Research* 52:102104. doi: 10.1016/j.algal.2020.102104

Gerardo, M. L., S. van den Hende, H. Vervaeren, T. Coward, and S. C. Skill. 2015. "Harvesting of microalgae within a biorefinery approach: A review of the developments and case studies from pilot-plants." *Algal Research* 11:248–262. doi: 10.1016/j.algal.2015.06.019

Griffiths, M., S. T. L. Harrison, M. Smit, and D. Maharajh. 2016. "Major Commercial Products from Micro- and Macroalgae." In *Algae Biotechnology*, edited by F. Bux and Y. Chisti, 269–300. Cham: Springer International Publishing.

Gu, D., K. Andreev, and M. E. Dupre. 2021. "Major trends in population growth around the world." *China CDC Weekly* 3 (28):604–613. doi: 10.46234/ccdcw2021.160

Ibrahim, T., N. A. S. Feisal, N. H. Kamaludin, W. Y. Cheah, V. How, A. Bhatnagar, Z. Ma, and P. L. Show. 2023. "Biological active metabolites from microalgae for healthcare and pharmaceutical industries: A comprehensive review." *Bioresource Technology* 372:128661. doi: 10.1016/j.biortech.2023.128661

Janssen, M., R. H. Wijffels, and M. J. Barbosa. 2022. "Microalgae based production of single-cell protein." *Current Opinion in Biotechnology* 75:102705. doi: 10.1016/j.copbio.2022.102705

Japar, A. S., M. S. Takriff, and N. H. M. Yasin. 2017. "Harvesting microalgal biomass and lipid extraction for potential biofuel production: A review." *Journal of Environmental Chemical Engineering* 5 (1):555–563. doi: 10.1016/j.jece.2016.12.016

Kapoore, R. V., E. E. Wood, and C. A. Llewellyn. 2021. "Algae biostimulants: A critical look at microalgal biostimulants for sustainable agricultural practices." *Biotechnology Advances* 49:107754. doi: 10.1016/j.biotechadv.2021.107754

Khalid, M. 2020. "Nanotechnology and chemical engineering as a tool to bioprocess microalgae for its applications in therapeutics and bioresource management." *Critical Reviews in Biotechnology* 40 (1):46–63. doi: 10.1080/07388551.2019.1680599

Kim, S. Y., Y. M. Kwon, K. W. Kim, and J. Y. Kim. 2021. "Exploring the potential of *Nannochloropsis* sp. extract for cosmeceutical applications." *Marine Drugs* 19 (12). doi: 10.3390/md19120690

Kumar, A., and J. S. Singh. 2020. "Microalgal Bio-Fertilizers." In *Handbook of Microalgae-Based Processes and Products*, edited by E. Jacob-Lopes, M. M. Maroneze, M. I. Queiroz, and L. Q. Zepka, 445–463. Academic Press.

Kumari, A., S. Pabbi, and A. Tyagi. 2023. "Recent advances in enhancing the production of long chain omega-3 fatty acids in microalgae." *Critical Reviews in Food Science and Nutrition* 26:1–19. doi: 10.1080/10408398.2023.2226720

La Bella, E., A. Baglieri, F. Fragalà, and I. Puglisi. 2022. "Multipurpose agricultural reuse of microalgae biomasses employed for the treatment of urban wastewater." *Agronomy* 12 (2):234. doi: 10.3390/agronomy12020234

Lee, S. Y., I. Khoiroh, D.-V. N. Vo, P. Senthil Kumar, and P. L. Show. 2020. "Techniques of lipid extraction from microalgae for biofuel production: A review." *Environmental Chemistry Letters* 19 (1):231–251. doi: 10.1007/s10311-020-01088-5

Leite, L. S., and L. A. Daniel. 2020. "Optimization of microalgae harvesting by sedimentation induced by high pH." *Water Science and Technology* 82 (6):1227–1236. doi: 10.2166/wst.2020.106

Leong, Y. K., and J. S. Chang. 2020. "Bioremediation of heavy metals using microalgae: Recent advances and mechanisms." *Bioresource Technology* 303:122886. doi: 10.1016/j.biortech.2020.122886

Liang, S. X. T., L. S. Wong, A. C. T. A. Dhanapal, and S. Djearamane. 2020. "Toxicity of metals and metallic nanoparticles on nutritional properties of microalgae." *Water, Air, & Soil Pollution* 231 (2):52. doi: 10.1007/s11270-020-4413-5

Loke Show, P. 2022. "Global market and economic analysis of microalgae technology: Status and perspectives." *Bioresource Technology* 357:127329. doi: 10.1016/j.biortech.2022.127329

Lucakova, S., I. Branyikova, and M. Hayes. 2022. "Microalgal proteins and bioactives for food, feed, and other applications." *Applied Sciences* 12 (9):4402. doi: 10.3390/app12094402

Maeda, Y., T. Yoshino, T. Matsunaga, M. Matsumoto, and T. Tanaka. 2018. "Marine microalgae for production of biofuels and chemicals." *Current Opinion in Biotechnology* 50:111–120. doi: 10.1016/j.copbio.2017.11.018

Marques, I. M., A. C. V. Oliveira, O. M. C. de Oliveira, E. A. Sales, and I. T. A. Moreira. 2021. "A photobioreactor using *Nannochloropsis oculata* marine microalgae for removal of polycyclic aromatic hydrocarbons and sorption of metals in produced water." *Chemosphere* 281:130775. doi: 10.1016/j.chemosphere.2021.130775

Matsunaga, T., H. Takeyama, H. Miyashita, and H. Yokouchi. 2005. "Marine Microalgae." In *Marine Biotechnology I*, edited by R. Ulber and Y. L. Gal, 165–188. Berlin, Heidelberg: Springer.

Matter, I. A., V. K. Hoang Bui, M. Jung, J. Y. Seo, Y.-E. Kim, Y.-C. Lee, and Y.-K. Oh. 2019. "Flocculation harvesting techniques for microalgae: A review." *Applied Sciences* 9 (15):3069. https://doi.org/10.3390/app9153069

Menegazzo, M. L., and G. G. Fonseca. 2019. "Biomass recovery and lipid extraction processes for microalgae biofuels production: A review." *Renewable and Sustainable Energy Reviews* 107:87–107. doi: https://doi.org/10.1016/j.rser.2019.01.064

Michalak, M., M. Pierzak, B. Kręcisz, and E. Suliga. 2021. "Bioactive compounds for skin health: A review." *Nutrients* 13 (1):203. doi: 10.3390/nu13010203

Molina Grima, E., E. H. Belarbi, F. G. Acien Fernandez, A. Robles Medina, and Y. Chisti. 2003. "Recovery of microalgal biomass and metabolites: Process options and economics." *Biotechnology Advances* 20 (7-8):491–515. doi: 10.1016/s0734-9750(02)00050-2

Napan, K., T. Christianson, K. Voie, and J. C. Quinn. 2015. "Quantitative assessment of microalgae biomass and lipid stability post-cultivation." *Frontiers in Energy Research* 3. doi: 10.3389/fenrg.2015.00015

Narala, R. R., S. Garg, K. K. Sharma, S. R. Thomas-Hall, M. Deme, Y. Li, and P. M. Schenk. 2016. "Comparison of microalgae cultivation in photobioreactor, open raceway pond, and a two-stage hybrid system." *Frontiers in Energy Research* 4. doi: 10.3389/fenrg.2016.00029

Norsker, N. H., M. J. Barbosa, M. H. Vermue, and R. H. Wijffels. 2011. "Microalgal production: A close look at the economics." *Biotechnology Advances* 29 (1):24–7. doi: 10.1016/j.biotechadv.2010.08.005

Nwoba, E. G., A. Vadiveloo, C. N. Ogbonna, B. E. Ubi, J. C. Ogbonna, and N. R. Moheimani. 2020. "Algal cultivation for treating wastewater in African developing countries: A review." *Clean – Soil, Air, Water* 48 (3):2000052. doi: 10.1002/clen.202000052

Okoro, V., U. Azimov, J. Munoz, H. H. Hernandez, and A. N. Phan. 2019. "Microalgae cultivation and harvesting: Growth performance and use of flocculants – A review." *Renewable and Sustainable Energy Reviews* 115:109364. doi: 10.1016/j.rser.2019.109364

Olguín, E. J., G. Sánchez-Galván, I. I. Arias-Olguín, F. J. Melo, R. E. González-Portela, L. Cruz, R. De Philippis, and A. Adessi. 2022. "Microalgae-based biorefineries: Challenges and future trends to produce carbohydrate enriched biomass, high-added value products and bioactive compounds." *Biology* 11 (8):1146. doi: 10.3390/biology11081146.

Peng, L., D. Fu, H. Chu, Z. Wang, and H. Qi. 2019. "Biofuel production from microalgae: A review." *Environmental Chemistry Letters* 18 (2):285–297. doi: 10.1007/s10311-019-00939-0

Pérez, L., J. L. Salgueiro, R. Maceiras, Á. Cancela, and Á. Sánchez. 2017. "An effective method for harvesting of marine microalgae: pH induced flocculation." *Biomass and Bioenergy* 97:20–26. doi: 10.1016/j.biombioe.2016.12.010

Plaza, B. M., C. Gómez-Serrano, F. G. Acién-Fernández, and S. Jimenez-Becker. 2018. "Effect of microalgae hydrolysate foliar application (*Arthrospira platensis* and *Scenedesmus* sp.) on Petunia × hybrida growth." *Journal of Applied Phycology* 30 (4):2359–2365. doi: 10.1007/s10811-018-1427-0

Prasad, R., S. K. Gupta, N. Shabnam, C. Y. B. Oliveira, A. K. Nema, F. A. Ansari, and F. Bux. 2021. "Role of microalgae in global CO_2 sequestration: Physiological mechanism, recent development, challenges, and future prospective." *Sustainability* 13 (23):13061. doi: 10.3390/su132313061

Pulz, O., and K. Scheibenbogen. 1998. "Photobioreactors: Design and Performance with Respect to Light Energy Input." In *Bioprocess and Algae Reactor Technology, Apoptosis*, 123–152. Berlin, Heidelberg: Springer.

Qin, L., M. A. Alam, and Z. Wang. 2019. "Open Pond Culture Systems and Photobioreactors for Microalgal Biofuel Production." In *Microalgae Biotechnology for Development of Biofuel and Wastewater Treatment*, edited by Md. A. Alam and Z. Wang, 45–74. Singapore: Springer.

Qiu, C., Y. He, Z. Huang, S. Li, J. Huang, M. Wang, and B. Chen. 2019. "Lipid extraction from wet *Nannochloropsis* biomass via enzyme-assisted three phase partitioning." *Bioresource Technology* 284:381–390. doi: 10.1016/j.biortech.2019.03.148

Rahman, K. M., and L. Melville. 2023. "Global Market Opportunities for Food and Feed Products from Microalgae." In *Handbook of Food and Feed from Microalgae*, edited by E. Jacob-Lopes, M. I. Queiroz, M. M. Maroneze and L. Q. Zepka, 593–602. Academic Press.

Ramesh Kumar, B., G. Deviram, T. Mathimani, P. A. Duc, and A. Pugazhendhi. 2019. "Microalgae as rich source of polyunsaturated fatty acids." *Biocatalysis and Agricultural Biotechnology* 17:583–588. doi: 10.1016/j.bcab.2019.01.017

Ramos-Romero, S., J. R. Torrella, T. Pagès, G. Viscor, and J. L. Torres. 2021. "Edible microalgae and their bioactive compounds in the prevention and treatment of metabolic alterations." *Nutrients* 13 (2):563. doi: 10.3390/nu13020563

Remize, M., Y. Brunel, J. L. Silva, J.-Y. Berthon, and E. Filaire. 2021. "Microalgae n-3 PUFAs production and use in food and feed industries." *Marine Drugs* 19 (2):113. doi: 10.3390/md19020113.

Renuka, N., A. Guldhe, R. Prasanna, P. Singh, and F. Bux. 2018. "Microalgae as multifunctional options in modern agriculture: Current trends, prospects and challenges." *Biotechnology Advances* 36 (4):1255–1273. doi: 10.1016/j.biotechadv.2018.04.004

Righini, H., and R. Roberti. 2019. "Algae and Cyanobacteria as Biocontrol Agents of Fungal Plant Pathogens." In *Plant Microbe Interface*, edited by A. Varma, S. Tripathi, and R. Prasad, 219–238. Cham: Springer International Publishing.

Roy, M., and K. Mohanty. 2019. "A comprehensive review on microalgal harvesting strategies: Current status and future prospects." *Algal Research* 44:101683. doi: 10.1016/j.algal.2019.101683

Ruiz Martinez, M. A., S. Peralta Galisteo, H. Castan, and M. E. Morales Hernandez. 2020. "Role of proteoglycans on skin ageing: A review." *International Journal of Cosmetic Science* 42 (6):529–535. doi: 10.1111/ics.12660.

Salim, S., R. Bosma, M. H. Vermue, and R. H. Wijffels. 2011. "Harvesting of microalgae by bio-flocculation." *Journal of Applied Phycology* 23 (5):849–855. doi: 10.1007/s10811-010-9591-x

Sanjeewa, K. K. A., I. P. S. Fernando, K. W. Samarakoon, H. H. C. Lakmal, E.-A. Kim, O. N. Kwon, M. G. Dilshara, J.-B. Lee, and Y.-J. Jeon. 2016. "Anti-inflammatory and anticancer activities of sterol rich fraction of cultured marine microalga *Nannochloropsis oculata*." *Algae* 31 (3):277–287. doi: 10.4490/algae.2016.31.6.29

Shokravi, Z., H. Shokravi, O. H. Chyuan, W. J. Lau, S. S. Koloor, M. Petrů, and A. F. Ismail. 2020. "Improving 'lipid Productivity' in microalgae by bilateral enhancement of biomass and lipid contents: A review." *Sustainability* 12 (21):9083. doi: 10.3390/su12219083.

Siddiki, S. Y. A., M. Mofijur, P. S. Kumar, S. F. Ahmed, A. Inayat, F. Kusumo, I. A. Badruddin, T. M. Yunus Khan, L. D. Nghiem, H. C. Ong, and T. M. I. Mahlia. 2022. "Microalgae biomass as a sustainable source for biofuel, biochemical and biobased value-added products: An integrated biorefinery concept." *Fuel* 307:121782. doi: 10.1016/j.fuel.2021.121782

Silva, S. C., I. C. F. R. Ferreira, M. M. Dias, and M. F. Barreiro. 2020. "Microalgae-derived pigments: A 10-year bibliometric review and industry and market trend analysis." *Molecules* 25 (15):3406. doi: 10.3390/molecules25153406

Singh, U. B., and A. S. Ahluwalia. 2012. "Microalgae: A promising tool for carbon sequestration." *Mitigation and Adaptation Strategies for Global Change* 18 (1):73–95. doi: 10.1007/s11027-012-9393-3.

Singh, R. N., and S. Sharma. 2012. "Development of suitable photobioreactor for algae production: A review." *Renewable and Sustainable Energy Reviews* 16 (4):2347–2353. doi: 10.1016/j.rser.2012.01.026

Spain, O., M. Plohn, and C. Funk. 2021. "The cell wall of green microalgae and its role in heavy metal removal." *Physiologia Plantarum* 173 (2):526–535. doi: 10.1111/ppl.13405

Spolaore, P., C. Joannis-Cassan, E. Duran, and A. Isambert. 2006. "Commercial applications of microalgae." *Journal of Bioscience and Bioengineering* 101 (2):87–96. doi: 10.1263/jbb.101.87

Suparmaniam, U., M. K. Lam, Y. Uemura, J. W. Lim, K. T. Lee, and S. H. Shuit. 2019. "Insights into the microalgae cultivation technology and harvesting process for biofuel production: A review." *Renewable and Sustainable Energy Reviews* 115:109361. doi: 10.1016/j.rser.2019.109361

Supriyanto, R. N., T. Ahamed, D. Mikihide, and M. M. Watanabe. 2018. "A decision tree approach to estimate the microalgae production in open raceway pond." *IOP Conference Series: Earth and Environmental Science* 209 (1):012050. doi: 10.1088/1755-1315/209/1/012050

Tocher, D. R., M. B. Betancor, M. Sprague, R. E. Olsen, and J. A. Napier. 2019. "Omega-3 long-chain polyunsaturated fatty acids, EPA and DHA: Bridging the gap between supply and demand." *Nutrients* 11 (1):89. doi: 10.3390/nu11010089

Touliabah, H. E., M. M. El-Sheekh, M. M. Ismail, and H. El-Kassas. 2022. "A review of microalgae- and cyanobacteria-based biodegradation of organic pollutants." *Molecules* 27 (3):1141. doi: 10.3390/molecules27031141

Ummalyma, S. B., E. Gnansounou, R. K. Sukumaran, R. Sindhu, A. Pandey, and D. Sahoo. 2017. "Bioflocculation: An alternative strategy for harvesting of microalgae – an overview." *Bioresource Technology* 242:227–235. doi: 10.1016/j.biortech.2017.02.097

van Dijk, M., T. Morley, M. L. Rau, and Y. Saghai. 2021. "A meta-analysis of projected global food demand and population at risk of hunger for the period 2010–2050." *Nature Food* 2 (7):494–501. doi: 10.1038/s43016-021-00322-9

Vasistha, S., A. Khanra, M. Clifford, and M. P. Rai. 2021. "Current advances in microalgae harvesting and lipid extraction processes for improved biodiesel production: A review." *Renewable and Sustainable Energy Reviews* 137:110498. doi: 10.1016/j.rser.2020.110498

Venkatesan, J., P. Manivasagan, and S.-K. Kim. 2015. "Marine Microalgae Biotechnology." In *Handbook of Marine Microalgae*, edited by S.-K. Kim, 1–9. Boston: Academic Press.

Vinokurova, G. V. 2023. "Algae of Lake Teletskoye ecotones." *Inland Water Biology* 16 (2):219–228. doi: 10.1134/s1995082923020220

Wang, Y., Y. Gong, L. Dai, M. Sommerfeld, C. Zhang, and Q. Hu. 2018. "Identification of harmful protozoa in outdoor cultivation of *Chlorella* and the use of ultra-sonication to control contamination." *Algal Research* 31:298–310. doi: 10.1016/j.algal.2018.02.002

Wang, S.-K., A. R. Stiles, C. Guo, and C.-Z. Liu. 2015. "Harvesting microalgae by magnetic separation: A review." *Algal Research* 9:178–185. doi: 10.1016/j.algal.2015.03.005

Xu, L., P. J. Weathers, X.-R. Xiong, and C.-Z. Liu. 2009. "Microalgal bioreactors: Challenges and opportunities." *Engineering in Life Sciences* 9 (3):178–189. doi: 10.1002/elsc.200800111

Xu, K., X. Zou, W. Chang, Y. Qu, and Y. Li. 2021. "Microalgae harvesting technique using ballasted flotation: A review." *Separation and Purification Technology* 276:119439. doi: 10.1016/j.seppur.2021.119439.

Yen, H.-W., I. C. Hu, C.-Y. Chen, D. Nagarajan, and J.-S. Chang. 2019. "Design of Photobioreactors for Algal Cultivation." In *Biofuels from Algae*, edited by A. Pandey, J.-S. C., Carlos Ricardo S., D.-J. Lee, and Y. Chisti, 225–256. Elsevier.

Zainith, S., G. Saxena, R. Kishor, and R. N. Bharagava. 2021. "Application of Microalgae in Industrial Effluent Treatment, Contaminants Removal, and Biodiesel Production: Opportunities, Challenges, and Future Prospects." In *Bioremediation for Environmental Sustainability*, edited by G. Saxena, V. Kumar, and M. P. Shah, 481–517. Elsevier.

Zhuang, D., N. He, K. S. Khoo, E. P. Ng, K. W. Chew, and T. C. Ling. 2022. "Application progress of bioactive compounds in microalgae on pharmaceutical and cosmetics." *Chemosphere* 291 (Pt 2):132932. doi: 10.1016/j.chemosphere.2021.132932

6 Marine Nematodes

6.1 BIODIVERSITY AND TAXONOMY OF MARINE NEMATODES

The morphology of free-living nematodes is simple. The gonad and intestine are enclosed by a sturdy and flexible body wall with ventral and dorsal longitudinal muscles. The pressurized, fluid-filled body cavity and body wall of nematodes work as an antagonist for muscle action and enable movement. This mode of locomotion limits morphological diversity in nematodes and the development/evolution of loco-motory appendages such as legs, fins, or wings is not reported from these tiny, useful marine organisms (Kiontke and Fitch 2013; Schratzberger et al. 2019). In general, the nematode mouth is a simple tube. However, in plant parasites and fungal feeders, the mouth is adorned with a piercing style or with frightening teeth that can cut, tear, or chew (Kiontke and Fitch 2013).

Nematodes are the most common metazoans in the biosphere, colonizing prac-tically all semi-aquatic and aquatic environments on the planet (Danovaro et al. 2023). They may be grown in lentic and lotic surface waters like streams and lakes, as well as on the seafloor. (Abebe, Decraemer, and De Ley 2007; Ptatscheck and Traunspurger 2020; Song et al. 2017). Specifically marine nematodes, in particular, have been recognized as a potentially diverse group, and the deep sea as a potentially complex and diverse ecosystem (Lambshead and Boucher 2003).

Free-living nematodes constitute the highest metazoan diversity in many benthic habitats (Merckx et al. 2009). A worldwide examination of marine and freshwater nematodes reveals distinct and nonrandom distribution patterns for marine nematodes. In general, marine nematode density decreases as water depth increases (Ptatscheck and Traunspurger 2020). According to the previous observa-tions for intertidal and upper subtidal (<50 m) zones, the average worldwide nema-tode density is 1,530 individuals/10 cm^2; for continental slope (1,000–2,500 m), the average nematode density is 430 individuals/10 cm^2; for lower slopes (2,500–3,500 m), the average nematode density is 360 individuals/10 cm^2; and for abyssal and hadal depths (>5,000 m), the average nematode density is 140 individuals/10 cm^2 (Udalov, Azovsky and Mokievsky 2005). Other than the depth, biogeochemical properties of the sediment are also found to affect the diversity and the composi-tion of marine nematode assemblages (Merckx et al. 2009). In general, these free-living marine nematodes have proven effective as indicators of biological health and ocean pollution for the last few decades; however, the bioindicator role of marine nematodes is not universally recognized (Hua et al. 2021). The primary use of nematodes has been as indicators of heavy metal and hydrocarbon pollution, with limited instances of their application to assess biological, environmental, or physical disturbances (Khandelwal et al. 2022; Salamun et al. 2012). Despite their status as excellent bioindicators due to their widespread distribution, abundance, and specific responses to environmental pollution at the genus and species levels, there remain challenges preventing their global utilization (Gyedu-Ababio et al.

DOI: 10.1201/9781003477365-6

1999; Ridall and Ingels 2021). In this chapter, potential applications of marine nematodes are discussed briefly.

6.2 APPLICATIONS OF MARINE NEMATODES

Beyond their traditional roles of marine nematodes as ecological indicators, marine nematodes are found to be useful in different sectors such as aquaculture and fisheries, pathogen management, environmental monitoring and bioindication, nutrient cycling applications, and bioremediation applications (Figure 6.1). In this section, these major applications are discussed in detail.

6.2.1 MARINE NEMATODES IN AQUACULTURE AND FISHERIES MANAGEMENT

Free-living marine nematodes are considered as a major organism in meiobenthic communities, where these nematodes serve as a link between microbial production and higher trophic levels (Weber and Traunspurger 2016). In modern-day aquaculture pond management, the role of nematodes extends beyond mere indicators of pond health; nematodes are integral contributors to the overall ecological dynamics. The presence and composition of nematode communities in pond sediments are pivotal components in ensuring the harmony of aquaculture ecosystems.

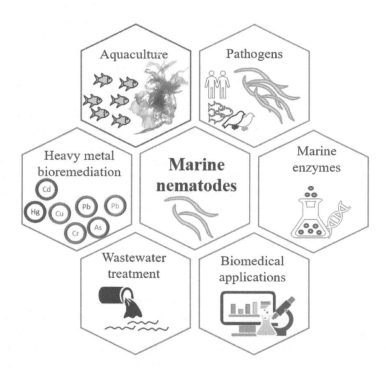

FIGURE 6.1 Potential applications of marine nematodes.

Nematodes, through their activities in sediment bioturbation, play a vital role in shaping the physical environment of aquaculture ponds (Chakraborty, Saha and Aditya 2022). The burrowing and feeding activities enhance oxygen diffusion into the sediment, reducing the risk of anaerobic conditions that can compromise water quality (Maciute et al. 2023). This bioturbation process fosters a healthy environment for the diverse microbial community, further contributing to nutrient cycling and the breakdown of organic matter.

Nutrient cycling within aquaculture ponds is a key aspect of pond management, and nematodes play a dual role (Chen et al. 2020; Farrant, Frank and Larsen 2021). As detritivores, they actively participate in the decomposition of organic material, releasing essential nutrients back into the pond water. The nematodes in the aquatic ponds can make proteins, lipids, and vitamins available to farmed animals, increasing growth and minimizing the commercial feed quantities required for aquaculture operations (Khanjani, Mohammadi and Emerenciano 2022). In general, aqua feed constitutes a key segment of the aquaculture cost; having beneficial nematodes in aquaculture ponds can be beneficial to the farmers in terms of cost reduction (Khanjani and Sharifinia 2020). This nutrient recycling supports the growth of not only target aquatic species but also other beneficial algae and other microorganisms; this further provides a nutrient-rich foundation for cultured organisms (Farrant, Frank and Larsen 2021; Weitere et al. 2018). Moreover, nematodes act as bioindicators not only for general pond health but also for specific water quality parameters (Biswal 2022). Nematode reactions to changes in dissolved oxygen and pH levels provide detailed insights into the general health of the aquatic environment. Nematode reactions to changes in dissolved oxygen and pH levels provide detailed insights into the general health of the aquatic environment (Dashfield et al. 2008). Aquaculturists can use this information to make informed decisions about aeration, water quality maintenance, and sustainable practices that align with the natural processes of the pond ecosystem.

In general, the integration of nematode dynamics into aquaculture pond management represents a holistic approach to sustainable aquaculture. Understanding the complex relationships between nematodes, nutrient cycling, and water quality allows for proactive management strategies. As aquaculture continues to evolve: incorporating the contributions of nematodes ensures not only the immediate success of the enterprise but also the long-term resilience and ecological balance of aquaculture systems in the face of environmental changes and human interventions.

6.2.2 MARINE NEMATODES AS ECOSYSTEM ARCHITECTS AND EMERGING PATHOGENS

In general, marine nematodes were previously considered as beneficial marine organisms due to their supportive role in nutrient cycling and ecological processes (Ptatscheck and Traunspurger 2020). However, some studies reported the potential emergence of marine nematodes as pathogens in aquatic ecosystems (Pereira and Gonzalez-Solis 2022). Specifically, the abundance as well as distribution of parasitic nematodes in marine macroorganisms such as fishes and clams are poorly studied in various geographic locations. Understanding the impact of environmental changes on nematode infection rates in these organisms requires immediate attention as these parasitic actions of marine nematodes can impact on global food security in the future (Oğuz et al. 2021). In this section, parasitic nature of marine nematodes and their impact on natural ecosystems and aquaculture environments are discussed briefly.

6.2.3 EFFECTS OF NEMATODES IN THE SEAFOOD INDUSTRY AND AQUACULTURE

Seafood consumption has increased worldwide, due to the proven health benefits associated with seafood consumption. In some countries, seafood has now surpassed the consumption of other proteins such as sheep and lamb (Hossen and Shamsi 2019). Consumption of raw and undercooked marine-based food items, in particular, has increased, and as a result, seafood-borne parasites are a public health problem in many parts of the world (Hossen and Shamsi 2019; Sumner et al. 2015). Many fish species are susceptible to infection by nematodes (phylum Chromadorea) (Moravec 2007). Endoparasitic nematodes in marine fish have an influence on fish health, fish quality for human consumption, and human health (Fiorenza et al. 2020; Truong et al. 2022). Moreover, a number of host species of marine nematodes have been widely investigated (*Anisakis* spp., *Pseudoterranova* spp., *Oncorhynchus* spp., and *Hippoglossus stenolepis*); the distribution and quantity of nematodes in other recognized hosts has received less attention (Oğuz et al. 2021). Efforts to discover and describe novel parasite species in fish hosts are critical not only for completing our biodiversity records but also for monitoring and mitigating disease concerns in fisheries and aquaculture in the face of global climate change (Poulin, Presswell and Jorge 2020).

In commercial aquaculture, diseases like parasitic nematode infections limit the further development of the aquaculture sector in many developing and developed countries. Nematodes infections have caused mortality, reduced growth and reproduction rates, and affected fitness of the culture fish stock, rendering infected fish unmarketable and eventually increasing the production cost (Amakali et al. 2023; Garza et al. 2019). Some fish infections are zoonotic (can be transmitted between animals and humans). Taken together, various studies globally attempted to address parasitic nematodes in aqua-cultured fishes and how they affect and/or influence the productivity of aquaculture organisms (Amakali et al. 2023). Besides their parasitic role, marine nematodes are considered as the most abundant prey of macro-megafauna and fish juveniles, thus playing a principal part in marine-based food webs (Danovaro et al. 2023).

6.2.4 MARINE NEMATODES IN ENVIRONMENTAL MONITORING AND BIOINDICATION

Ecological indicators can be used to identify the biological quality of a given ecosystem and to monitor the environmental changes in the particular ecosystem. Instead of chemical and physical metrics, indicator species (living organisms that are easily monitored and whose status reflects or predicts the condition(s) of the environment where they are found) can be used to monitor ecosystem health. In general, the health or abundance of indicator species reflects the environmental perturbation and provides the overall condition of the particular ecosystem (Siddig et al. 2016).

Nematodes have been identified as effective indicators of ecological conditions across different marine ecosystems, ranging from estuaries to deep-sea ecosystems (Hua et al. 2023; Liu et al. 2011). Their recognition as valuable bioindicators is attributed to a combination of characteristics: (i) widespread presence in high abundance and diversity nematode species; (ii) the short life cycle of most nematode species; (iii) the tolerance of certain species/genera to extreme conditions; (iv) respond rapidly to disturbance and enrichment; (v) transparent nature, ability to diagnose internal features without dissection. They can, therefore, be identified without biochemical procedures (Bongers and Ferris 1999; Fonseca and Gallucci 2016; Kennedy and Jacoby 1999). With these properties, nematodes can be used as bioindicators in different applications, including diverse ecosystem services, soil health assessments, level of pollution in marine ecosystems, monitoring plant diseases, managing parasites in aquatic animals, and safeguarding human health (Hagerbaumer et al. 2015; Nachev and Sures 2016; Yadav et al. 2022).

However, to use nematode-based indicators effectively for ecosystem-related quality parameter assessments requires fundamental knowledge of nematode biodiversity and functional patterns along with the drivers that generate those patterns (Franzo et al. 2022). For example, according to Sroczyńska et al. (2021), the nematode's functional responses differ from the taxonomy-based approach in relation to environmental variations (Sroczyńska et al. 2021). In addition, Santos, Cardoso, and Maria (2019) similarly noted that the evaluated biological indicators, assessed across various sites in Brazil and using the same nematode species, failed to provide clear definitive insights into the ecological well-being of the examined locations (Santos, Cardoso and Maria 2019).

6.2.5 Contribution of Marine Nematodes to Nutrient Cycling in Marine Ecosystems

Marine nematodes, often overshadowed by larger marine organisms, play a fundamental and often underappreciated role in nutrient cycling within marine ecosystems. These microscopic organisms contribute significantly to the breakdown of organic matter, the release of essential nutrients, and the overall health and productivity of marine environments (Moss 2017; Svetashev 2022). One of the primary contributions of marine nematodes to nutrient cycling lies in their role as decomposers (Milkereit et al. 2021). Usually, marine nematodes feed on a variety of organic materials, including benthic bacteria, fungi, microalgae, seagrass detritus microphytobenthos, and sediment particulate organic matter (Hwang et al. 2023). Through this feeding process, complex organic compounds are broken down into simpler forms, releasing critical nutrients such as nitrogen and phosphorus back into the sediment (di Montanara et al. 2022). Besides decomposition and nutrient enrichment activities, the digging and movement of marine nematodes within sediments play a crucial role in bioturbation, disrupting sediment layers (Murray, Meadows and Meadows 2002). This activity enhances oxygen penetration into the sediments, fostering aerobic conditions. Adequate oxygenation is essential for the microbial decomposition of organic matter, further promoting nutrient release in forms accessible to plants and other marine organisms (Murray, Meadows and Meadows 2002). In addition, marine nematodes engage in intricate interactions with microorganisms within sediments. These interactions influence the composition and activity of microbial communities, thereby impacting nutrient cycling processes (Murray, Meadows and Meadows 2002; Vanreusel et al. 2010; Yeates 2003). The feeding activities of nematodes regulate microbial populations, contributing to the delicate balance of nutrient transformations within marine environments.

Besides, the abundance and diversity of marine nematodes are responsive to environmental variables, including sediment composition, temperature, and food availability (Vanreusel et al. 2010). Changes in these variables can impact the activity and composition of nematode communities, subsequently influencing nutrient cycling rates in marine sediments. Understanding these responses is crucial for predicting how marine ecosystems may adapt to environmental changes.

6.2.6 BIOTECHNOLOGICAL POTENTIAL OF MARINE NEMATODES

Besides the parasitic and ecosystem engineering roles, marine nematodes have been recognized as a potential organism to use in biotechnological research and applications due to their intriguing attributes. Usually, nematodes display distinct biochemical and ecological properties that offer promise for a wide range of biotechnological applications.

Marine-based enzymes are a group of marine biomaterials with great interest among researchers as they are synthesized in harsh marine environmental conditions like wide-ranging pH, high pressure, high salinity, and wide temperature ranges. Therefore, marine-derived enzymes are capable of showing industrially important properties for a number of applications due to their unique composition (Ghattavi and Homaei 2023). Among the potential applications, these nematodes are useful organisms to search for industrially useful marine enzymes such as extremophilic enzymes, cold-adapted enzymes, biodegradative enzymes, bioprospecting for novel enzymes, applications in biocatalysis, and industrial and biomedical uses. However, there are no detailed scholarly articles available for enzymes isolated from marine nematodes for the above activities. Thus, it is of utmost importance to expand the current research focus to identify enzymes from nematodes to use in the aforementioned applications.

6.2.7 BIOREMEDIATION POTENTIAL OF MARINE NEMATODES IN POLLUTED ENVIRONMENTS

In many parts of the world, the oceans have traditionally been identified as natural dump yards for garbage disposal. Thus, this ultimately leads to the increased levels of pollution we are observing today and the destruction of marine ecosystems.

Sources of marine pollution range from microplastics and pharmaceutical compounds to abiotic factors, such as sediment and runoff nutrients. With the increasing populations and resource consumption, the amount of potential pollution released to marine ecosystems has also been found to increase recently (Barboza et al. 2019; Willis et al. 2022; Zhou et al. 2022). The alarming increase in environmental pollution has prompted a quest for innovative and sustainable solutions, and marine nematodes are emerging as potential contributors to bioremediation efforts. As marine nematodes are abundant in marine sediments, they exhibit unique characteristics that position them as promising candidates for mitigating pollution in marine ecosystems (Table 6.1).

One of the major bioremediation applications of marine nematodes is cleaning up heavy metal-polluted marine environments. According to earlier studies, some marine nematode species show promise in metal bioremediation, particularly in environments contaminated with heavy metals. Their ability to sequester, transform, or tolerate metals contributes to the overall remediation of metal-polluted sediments (Sharma et al. 2022). For instance, a study carried out by Sedaghat et al. (2022) attempted to compare heavy metal concentrations in *Psettodes erumei* as host fish and larvae of *Hysterothylacium* spp. as its parasite. According to the results, the parasites had a significantly higher level of Ni, Fe, Zn, and Cu than the muscles of the host fish (Sedaghat et al. 2022). In addition, Baghdadi et al. (2023) also reported the metal-iron accumulation properties of *Hysterothylacium reliquens* isolated from king soldier bream *Argyrops spinifer*. Baghdadi et al. (2023) found results similar to Sedaghat et al. (2022), indicating that the larval stage of nematodes has the capacity to accumulate quantities of various metals, including Fe, Cu, Cd, and Pb (Baghdadi et al. 2023). In addition, Danovaro et al.

TABLE 6.1
Potential Bioremediation Applications Reported from Marine Nematodes

No.	Name	Potential Application	Reference
1	*Hysterothylacium* spp.	Heavy metal bioremediation	Sedaghat et al. (2022)
2	*Hysterothylacium reliquens*	Heavy metal bioremediation	Baghdadi et al. (2023)
3	Marine nematodes (species not specified)	Heavy metal bioremediation	Danovaro et al. (2023)
4	*Rhabditis (Pellioditis) marina*	Barium bioremediation	Lira et al. (2011)
5	*Bathylaimus capacosus* and *Bathylaimus tenuicaudatus*	Chromium bioremediation	Boufahja et al. (2011)
6	*Spirinia parasitifera*	Polycyclic aromatic hydrocarbon (phenanthrene) bioremediation	Louati et al. (2015)
7	*Marylynnia* spp.	Crude oil contamination bioremediation	Lv et al. (2011)
8	*Spirinia gerlachi*	Mineral lubricants bioremediation	Beyrem et al. (2010)
9	*Terschellingia longicaudata*	Synthetic lubricant bioremediation	Beyrem et al. (2010)
10	*Marylynnia stekhoveni*	Mercury bioremediation	Hermi et al. (2009)

(2023) also provide evidence of the bioaccumulation of heavy metals in marine nematodes. According to the authors, nematodes living in contaminated sediments had higher concentrations of Cr, Cd, and As in their tissues than those living in control sediments (Danovaro et al. 2023). These results suggest the potential of marine nematodes to be used in heavy metal bioremediation applications in the future. Lira et al. (2011) also reported the effects of Ba and Cd on the population development of *Rhabditis* (*Pellioditis*) *marina* marine nematodes. According to the authors, tested nematodes were able to tolerate Ba at concentrations up to 300 ppm, suggesting the potential of *R.* (*P.*) *marina* to be used in heavy metal bioremediation applications (Lira et al. 2011). Additionally, Boufahja et al. (2011) also reported high growth rates of two *Bathylaimus* species (*Bathylaimus capacosus* and *Bathylaimus tenuicaudatus*) in chromium-contaminated environments (Boufahja et al. 2011).

Some marine nematodes were also found to show promising potential against polycyclic aromatic hydrocarbon pollution. Previously, Louati et al. (2015) attempted to analyze benthic nematode reactions to polycyclic aromatic hydrocarbon (PAH) pollution. Additionally, the authors also attempted to examine bioremediation potential and efficacy in PAH degradation and impacts of PAH on nematodes. According to the results, *Spirinia parasitifera* emerged as the predominant species, constituting 70% of the relative abundance and displaying characteristics of an 'opportunistic' species in response to PAH contamination. Conversely, *Neochromadora peocilosoma* and *Oncholaimus campylocercoides* exhibited substantial inhibition in PAH-contaminated media. Biostimulation, involving the addition of a mineral salt medium, and bioaugmentation, entailing the inoculation of a hydrocarbonoclastic bacterium, served as bioremediation techniques. According to the authors, these approaches effectively improved the degradation of all tested PAHs under the tested conditions. Moreover, the authors reported 96% degradation for phenanthrene at the tested conditions (Louati et al. 2015).

In a similar study, Lv et al. (2011) attempted to evaluate the responses of marine nematode assemblages to crude oil contamination in the intertidal zone of Bathing Beach, Dalian. According to the results, nematode abundance, diversity, and species richness were found to decrease significantly with increasing levels of crude oil contamination. However, according to the authors, the abundance of *Marylynnia* sp. was found to increase slightly (Lv et al. 2011). These findings indicate that *Marylynnia* sp. has the potential to be employed in bioremediation applications where crude oil pollution has been observed. The effects of two lubricant oils on marine nematode communities in a laboratory microcosm study were reported by Beyrem et al. (2010). According to the authors, *Spirinia gerlachi* was found to increase in numbers in treatments involving mineral lubricants (both 'clean' and used), while it was completely eliminated in all treatments involving synthetic lubricants. This categorizes the species as 'resistant' to mineral oil contamination but intolerant to contamination from synthetic lubricants. On the other hand, *Terschellingia longicaudata* demonstrated an increase only in treatments with synthetic lubricants (both 'clean' and used), suggesting its categorization as a 'synthetic oil–resistant' species (Beyrem et al. 2010).

6.3 FUTURE DIRECTIONS

Marine nematodes are undervalued marine bioresource with potential applications in a number of industries, including sectors such as aquaculture and fisheries, pathogen management, environmental monitoring and bioindication, nutrient cycling applications, and bioremediation applications. However, further studies are required to explore the full potential of marine nematodes: specifically, areas like nematode enzymes and heavy metal absorption. The integration of marine nematode-derived enzymes and their heavy metal bioremediation properties into various industries holds the potential to introduce eco-friendly and efficient solutions.

REFERENCES

Abebe, E., W. Decraemer, and P. De Ley. 2007. "Global Diversity of Nematodes (Nematoda) in Freshwater." In *Freshwater Animal Diversity Assessment*, edited by E. V. Balian, C. Lévêque, H. Segers, and K. Martens, 67–78. Dordrecht, The Netherlands: Springer.

Amakali, A. M., A. Halajian, M. R. Wilhelm, M. Tjipute, and W. J. Luus-Powell. 2023. "Some Significant Parasites in Aquaculture and Their Potential Impact on the Development of Aquaculture in Africa." In *Emerging Sustainable Aquaculture Innovations in Africa*, edited by N. N. Gabriel, E. Omoregie, and K. P. Abasubong, 505–523. Singapore: Springer Nature.

Baghdadi, H. B., R. Abdel-Gaber, S. Al Quraishy, M. M. Abou Hadied, T. Al-Otaibi, M. F. Elkhadragy, E. M. Al-Shaebi, and M. Dkhil. 2023. "Metal accumulation capacity of raphidascaridid nematode, *Hysterothylacium reliquens*, infecting the king soldier bream (*Argyrops spinifer*)." *Journal of King Saud University: Science* 35 (4):102635. doi: 10.1016/j.jksus.2023.102635

Barboza, L. G. A., A. Cózar, B. C. G. Gimenez, T. L. Barros, P. J. Kershaw, and L. Guilhermino. 2019. "Macroplastics Pollution in the Marine Environment." In *World Seas: An Environmental Evaluation*, edited by C. Sheppard, 305–328. Academic Press.

Beyrem, H., H. Louati, N. Essid, P. Aissa, and E. Mahmoudi. 2010. "Effects of two lubricant oils on marine nematode assemblages in a laboratory microcosm experiment." *Marine Environmental Research* 69 (4):248–53. doi: 10.1016/j.marenvres.2009.10.018

Biswal, D. 2022. "Nematodes as ghosts of land use past: Elucidating the roles of soil nematode community studies as indicators of soil health and land management practices." *Applied Biochemistry and Biotechnology* 194 (5):2357–2417. doi: 10.1007/s12010-022-03808-9

Bongers, T., and H. Ferris. 1999. "Nematode community structure as a bioindicator in environmental monitoring." *Trends in Ecology & Evolution* 14 (6):224–228. doi: 10.1016/s0169-5347(98)01583-3

Boufahja, F., A. Hedfi, J. Amorri, P. Aissa, H. Beyrem, and E. Mahmoudi. 2011. "An assessment of the impact of chromium-amended sediment on a marine nematode assemblage using microcosm bioassays." *Biological Trace Element Research* 142 (2):242–55. doi: 10.1007/s12011-010-8762-6

Chakraborty, A., G. K. Saha, and G. Aditya. 2022. "Macroinvertebrates as engineers for bioturbation in freshwater ecosystem." *Environmental Science and Pollution Research* 29 (43):64447–64468. doi: 10.1007/s11356-022-22030-y

Chen, X., X. Chen, Y. Zhao, H. Zhou, X. Xiong, and C. Wu. 2020. "Effects of microplastic biofilms on nutrient cycling in simulated freshwater systems." *Science of the Total Environment* 719:137276. doi: 10.1016/j.scitotenv.2020.137276

Danovaro, R., A. Cocozza di Montanara, C. Corinaldesi, A. Dell'Anno, S. Illuminati, T. J. Willis, and C. Gambi. 2023. "Bioaccumulation and biomagnification of heavy metals in marine micro-predators." *Communications Biology* 6 (1):1206. doi: 10.1038/s42003-023-05539-x

Dashfield, S. L., P. J. Somerfield, S. Widdicombe, M. C. Austen, and M. Nimmo. 2008. "Impacts of ocean acidification and burrowing urchins on within-sediment pH profiles and subtidal nematode communities." *Journal of Experimental Marine Biology and Ecology* 365 (1):46–52. doi: 10.1016/j.jembe.2008.07.039

di Montanara, A. C., E. Baldrighi, A. Franzo, L. Catani, E. Grassi, R. Sandulli, and F. Semprucci. 2022. "Free-living nematodes research: State of the art, prospects, and future directions – a bibliometric analysis approach." *Ecological Informatics* 72:101891. doi: 10.1016/j.ecoinf.2022.101891

Farrant, D. N., K. L. Frank, and A. E. Larsen. 2021. "Reuse and recycle: Integrating aquaculture and agricultural systems to increase production and reduce nutrient pollution." *Science of the Total Environment* 785:146859. doi: 10.1016/j.scitotenv.2021.146859

Fiorenza, E. A., C. A. Wendt, K. A. Dobkowski, T. L. King, M. Pappaionou, P. Rabinowitz, J. F. Samhouri, and C. L. Wood. 2020. "It's a wormy world: Meta-analysis reveals several decades of change in the global abundance of the parasitic nematodes *Anisakis* spp. and *Pseudoterranova* spp. in marine fishes and invertebrates." *Global Change Biology* 26 (5):2854–2866. doi: 10.1111/gcb.15048

Fonseca, G., and F. Gallucci. 2016. "The need of hypothesis-driven designs and conceptual models in impact assessment studies: An example from the free-living marine nematodes." *Ecological Indicators* 71:79–86. doi: 10.1016/j.ecolind.2016.06.051

Franzo, A., E. Baldrighi, E. Grassi, M. Grego, M. Balsamo, M. Basili, and F. Semprucci. 2022. "Free-living nematodes of Mediterranean ports: A mandatory contribution for their use in ecological quality assessment." *Marine Pollution Bulletin* 180:113814. doi: 10.1016/j.marpolbul.2022.113814

Garza, M., C. V. Mohan, M. Rahman, B. Wieland, and B. Hasler. 2019. "The role of infectious disease impact in informing decision-making for animal health management in aquaculture systems in Bangladesh." *Preventive Veterinary Medicine* 167:202–213. doi: 10.1016/j.prevetmed.2018.03.004

Ghattavi, S., and A. Homaei. 2023. "Marine enzymes: Classification and application in various industries." *International Journal of Biological Macromolecules* 230:123136. doi: 10.1016/j.ijbiomac.2023.123136

Gyedu-Ababio, T. K., J. P. Furstenberg, D. Baird, and A. Vanreusel. 1999. "Nematodes as indicators of pollution: A case study from the Swartkops River system, South Africa." *Hydrobiologia* 397 (0):155–169. doi: 10.1023/a:1003617825985

Hagerbaumer, A., S. Hoss, P. Heininger, and W. Traunspurger. 2015. "Experimental studies with nematodes in ecotoxicology: An overview." *Journal of Nematology* 47 (1):11–27.

Hermi, M., E. Mahmoudi, H. Beyrem, P. Aissa, and N. Essid. 2009. "Responses of a free-living marine nematode community to mercury contamination: Results from microcosm experiments." *Archives of Environmental Contamination and Toxicology* 56 (3): 426–33. doi: 10.1007/s00244-008-9217-3

Hossen, M. S., and S. Shamsi. 2019. "Zoonotic nematode parasites infecting selected edible fish in new South Wales, Australia." *International Journal of Food Microbiology* 308:108306. doi: 10.1016/j.ijfoodmicro.2019.108306

Hua, E., L. He, Z. Zhang, C. Cui, and X. Liu. 2023. "Bioassessment of environmental quality based on taxonomic and functional traits of marine nematodes in the Bohai Sea, China." *Marine Pollution Bulletin* 190:114884. doi: 10.1016/j.marpolbul.2023.114884

Hua, E., Y. Zhu, D. Huang, and X. Liu. 2021. "Are free-living nematodes effective environmental quality indicators? Insights from Bohai Bay, China." *Ecological Indicators* 127:107756. doi: 10.1016/j.ecolind.2021.107756

Hwang, J.-M., H.-G. Kim, H. Kim, C.-H. Hwang, and C.-W. Oh. 2023. "Meiofaunal assemblages associated with macroalgal detritus decomposition." *Regional Studies in Marine Science* 68:103285. doi: 10.1016/j.rsma.2023.103285

Kennedy, A. D., and C. A. Jacoby. 1999. "Biological indicators of marine environmental health: Meiofauna – a neglected benthic component?" *Environmental Monitoring and Assessment* 54 (1):47–68. doi: 10.1023/a:1005854731889

Khandelwal, G., V. Chaudhary, R. Iyer, and A. Dwivedi. 2022. "Soil Bacteria and Nematodes for Bioremediation and Amelioration of Polluted Soil." In *Microbial and Biotechnological Interventions in Bioremediation and Phytoremediation*, edited by J. A. Malik, 57–79. Cham: Springer International Publishing.

Khanjani, M. H., A. Mohammadi, and M. G. C. Emerenciano. 2022. "Microorganisms in biofloc aquaculture system." *Aquaculture Reports* 26:101300. doi: 10.1016/j.aqrep.2022.101300

Khanjani, M. H., and M. Sharifinia. 2020. "Biofloc technology as a promising tool to improve aquaculture production." *Reviews in Aquaculture* 12 (3):1836–1850. doi: 10.1111/raq.12412

Kiontke, K., and D. H. Fitch. 2013. "Nematodes." *Current Biology* 23 (19):R862–R864. doi: 10.1016/j.cub.2013.08.009

Lambshead, P. J. D., and G. Boucher. 2003. "Marine nematode deep-sea biodiversity: Hyperdiverse or hype?" *Journal of Biogeography* 30 (4):475–485. doi: 10.1046/j.1365-2699.2003.00843.x

Lira, V. F., G. A. Santos, S. Derycke, M. E. Larrazabal, V. G. Fonseca-Genevois, and T. Moens. 2011. "Effects of barium and cadmium on the population development of the marine nematode *Rhabditis (Pellioditis) marina*." *Marine Environmental Research* 72 (4):151–159. doi: 10.1016/j.marenvres.2011.07.003

Liu, X. S., W. Z. Xu, S. G. Cheung, and P. K. Shin. 2011. "Marine meiobenthic and nematode community structure in Victoria Harbour, Hong Kong upon recovery from sewage pollution." *Marine Pollution Bulletin* 63 (5–12):318–325. doi: 10.1016/j.marpolbul.2011.03.027

Louati, H., O. B. Said, A. Soltani, C. Cravo-Laureau, R. Duran, P. Aissa, E. Mahmoudi, and O. Pringault. 2015. "Responses of a free-living benthic marine nematode community to bioremediation of a PAH mixture." *Environmental Science and Pollution Research* 22 (20):15307–15318. doi: 10.1007/s11356-014-3343-4

Lv, Y., W. Zhang, Y. Gao, S. Ning, and B. Yang. 2011. "Preliminary study on responses of marine nematode community to crude oil contamination in intertidal zone of bathing beach, Dalian." *Marine Pollution Bulletin* 62 (12):2700–2706. doi: 10.1016/j.marpolbul.2011.09.018

Maciute, A., O. Holovachov, R. N. Glud, E. Broman, P. Berg, F. J. A. Nascimento, and S. Bonaglia. 2023. "Reconciling the importance of meiofauna respiration for oxygen demand in muddy coastal sediments." *Limnology and Oceanography* 68 (8): 1895–1905. doi: 10.1002/lno.12393

Merckx, B., P. Goethals, M. Steyaert, A. Vanreusel, M. Vincx, and J. Vanaverbeke. 2009. "Predictability of marine nematode biodiversity." *Ecological Modelling* 220 (11): 1449–1458. doi: 10.1016/j.ecolmodel.2009.03.016

Milkereit, J., D. Geisseler, P. Lazicki, M. L. Settles, B. P. Durbin-Johnson, and A. Hodson. 2021. "Interactions between nitrogen availability, bacterial communities, and nematode indicators of soil food web function in response to organic amendments." *Applied Soil Ecology* 157:103767. doi: 10.1016/j.apsoil.2020.103767

Moravec, F. 2007. "Nematode parasites of fishes: Recent advances and problems of their research." *Parassitologia* 49 (3):155–60.

Moss, B. 2017. "Marine reptiles, birds and mammals and nutrient transfers among the seas and the land: An appraisal of current knowledge." *Journal of Experimental Marine Biology and Ecology* 492:63–80. doi: 10.1016/j.jembe.2017.01.018

Murray, J. M. H., A. Meadows, and P. S. Meadows. 2002. "Biogeomorphological implications of microscale interactions between sediment geotechnics and marine benthos: A review." *Geomorphology* 47 (1):15–30. doi: 10.1016/s0169-555x(02)00138-1

Nachev, M., and B. Sures. 2016. "Environmental parasitology: Parasites as accumulation bioindicators in the marine environment." *Journal of Sea Research* 113:45–50. doi: 10.1016/j.seares.2015.06.005

Oğuz, M. C., A. M. Campbell, S. P. Bennett, and M. C. Belk. 2021. "Nematode parasites of rockfish (*Sebastes* spp.) and cod (*Gadus* spp.) from waters near Kodiak Island Alaska, USA." *Diversity* 13 (9):436. doi: 10.3390/d13090436

Pereira, F. B., and D. Gonzalez-Solis. 2022. "Review of the parasitic nematodes of marine fishes from off the American continent." *Parasitology* 149 (14):1928–1941. doi: 10.1017/S0031182022001287

Poulin, R., B. Presswell, and F. Jorge. 2020. "The state of fish parasite discovery and taxonomy: A critical assessment and a look forward." *International Journal for Parasitology* 50 (10–11):733–742. doi: 10.1016/j.ijpara.2019.12.009

Ptatscheck, C., and W. Traunspurger. 2020. "The ability to get everywhere: Dispersal modes of free-living, aquatic nematodes." *Hydrobiologia* 847 (17):3519–3547. doi: 10.1007/s10750-020-04373-0

Ridall, A., and J. Ingels. 2021. "Suitability of free-living marine nematodes as bioindicators: Status and future considerations." *Frontiers in Marine Science* 8. doi: 10.3389/fmars.2021.685327

Salamun, P., M. Renco, E. Kucanova, T. Brazova, I. Papajova, D. Miklisova, and V. Hanzelova. 2012. "Nematodes as bioindicators of soil degradation due to heavy metals." *Ecotoxicology* 21 (8):2319–2330. doi: 10.1007/s10646-012-0988-y

Santos, G. H. C., R. S. Cardoso, and T. F. Maria. 2019. "Bioindicators or sediment relationships: Evaluating ecological responses from sandy beach nematodes." *Estuarine, Coastal and Shelf Science* 224:217–227. doi: 10.1016/j.ecss.2019.04.035

Schratzberger, M., M. Holterman, D. van Oevelen, and J. Helder. 2019. "A Worm's world: Ecological flexibility pays off for free-living nematodes in sediments and soils." *Bioscience* 69 (11):867–876. doi: 10.1093/biosci/biz086

Sedaghat, B., S. M. Sadjjadi, G. Mohebbi, and M. Rayani. 2022. "Metal uptake in *Psettodes erumei* and *Hysterothylacium* spp. larvae in the Persian Gulf: Evaluation of larvae as bio-indicator." *Journal of Parasitic Diseases* 46 (2):421–428. doi: 10.1007/s12639-021-01462-2

Sharma, I., H. Pandey, K. Thakur, and D. Pandey. 2022. "Phytochelatins and Their Application in Bioremediation." In *Microbial and Biotechnological Interventions in Bioremediation and Phytoremediation*, edited by J. A. Malik, 81–109. Cham: Springer International Publishing.

Siddig, A. A. H., A. M. Ellison, A. Ochs, C. Villar-Leeman, and M. K. Lau. 2016. "How do ecologists select and use indicator species to monitor ecological change? Insights from 14 years of publication in ecological indicators." *Ecological Indicators* 60:223–230. doi: 10.1016/j.ecolind.2015.06.036

Song, D., K. Pan, A. Tariq, F. Sun, Z. Li, X. Sun, L. Zhang, O. A. Olusanya, and X. Wu. 2017. "Large-scale patterns of distribution and diversity of terrestrial nematodes." *Applied Soil Ecology* 114:161–169. doi: 10.1016/j.apsoil.2017.02.013

Sroczyńska, K., P. Chainho, S. Vieira, and H. Adão. 2021. "What makes a better indicator? Taxonomic vs functional response of nematodes to estuarine gradient." *Ecological Indicators* 121:107113. doi: 10.1016/j.ecolind.2020.107113

Sumner, J., S. Antananawat, A. Kiermeier, C. McLeod, and S. Shamsi. 2015. "Raw fish consumption in Australia: How safe is it?" *Food Australia* 67 (3):24–26.

Svetashev, V. I. 2022. "Investigation of deep-sea ecosystems using marker fatty acids: Sources of essential polyunsaturated fatty acids in abyssal megafauna." *Marine Drugs* 20 (1). doi: 10.3390/md20010017

Truong, V. T., H. T. T. Ngo, T. Q. Bui, H. W. Palm, and R. A. Bray. 2022. "Marine fish parasites of Vietnam: A comprehensive review and updated list of species, hosts, and zoogeographical distribution." *Parasite* 29:36. doi: 10.1051/parasite/2022033

Udalov, A. A., A. I. Azovsky, and V. O. Mokievsky. 2005. "Depth-related pattern in nematode size: What does the depth itself really mean?" *Progress in Oceanography* 67 (1–2): 1–23. doi: 10.1016/j.pocean.2005.02.020

Vanreusel, A., A. De Groote, S. Gollner, and M. Bright. 2010. "Ecology and biogeography of free-living nematodes associated with chemosynthetic environments in the deep sea: A review." *PLoS One* 5 (8):e12449. doi: 10.1371/journal.pone.0012449

Weber, S., and W. Traunspurger. 2016. "Effects of juvenile fish predation (*Cyprinus carpio* L.) on the composition and diversity of free-living freshwater nematode assemblages." *Nematology* 18 (1):39–52. doi: 10.1163/15685411-00002941

Weitere, M., M. Erken, N. Majdi, H. Arndt, H. Norf, M. Reinshagen, W. Traunspurger, A. Walterscheid, and J. K. Wey. 2018. "The food web perspective on aquatic biofilms." *Ecological Monographs* 88 (4):543–559. doi: 10.1002/ecm.1315

Willis, K. A., C. Serra-Goncalves, K. Richardson, Q. A. Schuyler, H. Pedersen, K. Anderson, J. S. Stark, J. Vince, B. D. Hardesty, C. Wilcox, B. F. Nowak, J. L. Lavers, J. M. Semmens, D. Greeno, C. MacLeod, N. P. O. Frederiksen, and P. S. Puskic. 2022. "Cleaner seas: Reducing marine pollution." *Reviews in Fish Biology and Fisheries* 32 (1):145–160. doi: 10.1007/s11160-021-09674-8

Yadav, A., N. Kapoor, A. Arif, and S. K. Malhotra. 2022. "Energy dispersive X-ray microanalysis in conjunction with scanning electron micrography to establish nematodes as bioindicators in marine fish environment." *Journal of Parasitic Diseases* 46 (3): 664–671. doi: 10.1007/s12639-022-01480-8

Yeates, G. W. 2003. "Nematodes as soil indicators: Functional and biodiversity aspects." *Biology and Fertility of Soils* 37 (4):199–210. doi: 10.1007/s00374-003-0586-5

Zhou, Q., S. Wang, J. Liu, X. Hu, Y. Liu, Y. He, X. He, and X. Wu. 2022. "Geological evolution of offshore pollution and its long-term potential impacts on marine ecosystems." *Geoscience Frontiers* 13 (5):101427. doi: 10.1016/j.gsf.2022.101427

7 Applications of Marine Bioresources

7.1 INTRODUCTION

Marine bioresources, comprising a diverse array of organisms and compounds from the ocean, find applications across multiple domains (Imhoff, Labes and Wiese 2011). Mostly these marine resources are rich reservoirs of bioactive compounds, showing promise in the development of antibiotics, antivirals, anti-inflammatory, cardiovascular protection, and anticancer drugs (Daniotti and Re 2021; Sanjeewa et al. 2023). The inclusion of bioactive compounds in nutraceuticals and functional foods such as omega-3 fatty acid-rich fish oil and sulfated polysaccharides like fucoidan supports human health, promoting cardiovascular and cognitive well-being (Innes and Calder 2020; Shen et al. 2018). Moreover, cosmetics benefit from compounds sourced from marine organisms, such as fish extracts, seaweeds, microalgae, and other microorganisms, found to demonstrate ideal properties to develop into active ingredients in cosmeceuticals (Sotelo et al. 2021). Beyond the direct applications, marine bioresources play roles in environmental monitoring as bioindicators, contribute to bioenergy production, aid in bioremediation efforts, and inspire materials science innovations. Moreover, they serve as valuable subjects for research, education, and even traditional medicine in certain cultures. In this chapter, the author summarizes the potential applications of marine bioresources.

7.2 APPLICATIONS OF MARINE PLANT RESOURCES IN NEW PLANT FERTILIZER PRODUCTS

The exploration of marine plant resources for the development of innovative plant fertilizer products represents a compelling avenue in agricultural research. With the increasing demand for sustainable and environment-friendly agricultural practices, harnessing the unique properties of marine-derived compounds has gained significant attention (Nayaka, Toppo and Verma 2017; Nedumaran 2017; Pereira and Cotas 2019). Beyond merely supplying essential nutrients, these marine-based fertilizers have the capacity to offer numerous advantages, such as seed germination stimulation, plant health improvement, and plant development enhancements such as better water and nutrient uptake, salinity tolerance, shoot and root elongation, cold and biocontrol and resistance to phytopathogenic organisms, polluted soil cleanup, and fertilizing (Nabti, Jha and Hartmann 2016) (Figure 7.1). In addition, compared to organic fertilizers, marine-derived fertilizers are nontoxic and hazardous, biodegradable, and eco-friendly to use (Cook et al. 2018; Raghunandan et al. 2019). This section mainly focused on specific applications of marine plant resources in new

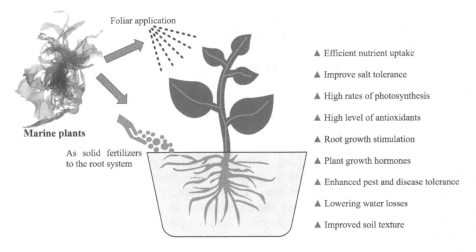

Foliar application

▲ Efficient nutrient uptake

▲ Improve salt tolerance

▲ High rates of photosynthesis

▲ High level of antioxidants

Marine plants

As solid fertilizers
to the root system

▲ Root growth stimulation

▲ Plant growth hormones

▲ Enhanced pest and disease tolerance

▲ Lowering water losses

▲ Improved soil texture

FIGURE 7.1   Advantages of marine-based plant fertilizers.

plant fertilizer products, highlighting their potential to revolutionize agricultural practices and contribute to a more resilient and ecologically responsible approach to farming.

7.2.1 BIOSTIMULATION PROPERTIES REPORTED FROM MARINE PLANT FERTILIZERS

Biostimulants enhance plant growth during the crop cycle via boosting plant metabolic efficiency, quality of the crops, and quick recovery after abiotic and biotic challenges. Marine macroalgae bio-extracts operate as biostimulants for land plants, with these effects attributed to plant growth regulators or phytohormones, as well as chemical compounds that can boost tolerance to a variety of biotic and abiotic challenges (Di Mola et al. 2019; Melo et al. 2020). Several studies highlighted the biostimulation properties of marine organisms. In a recent study, Melo et al. (2020) reported the biostimulation potential of extracts prepared from *Kappaphycus alvarezii* and *Sargasum vulgare* on pepper grown in greenhouse conditions (Melo et al. 2020).In addition, Ertani et al. (2018) reported the biostimulation potential of extracts prepared from *Ascophyllum nodosum* spp. and *Laminaria* and as biostimulants in *Zea mays* (Ertani et al. 2018). Other than seaweeds, microalgal polysaccharides were also found to possess promising biostimulation properties (Chanda, Merghoub and El Arroussi 2019). In addition, Chiaiese et al. (2018) also summarized the potential of marine plant resources to use as biostimulants (Chiaiese et al. 2018). Furthermore, Rathore et al. (2009) reported that seaweed extract derived from *K. alvarezii* is effective in enhancing various growth indicators, such as the height of the crops, plant density, the number of pods per plant, the number of grains per pod, the number of branches per plant, and test weight (Rathore et al. 2009).

7.2.2 Alleviation of Soil Salinity Stress

Salinity is a major issue in agricultural lands, negatively affecting crop growth. In general, it causes significant crop yield losses worldwide (AbdElgawad et al. 2016). In addition to their biostimulation properties, marine-based fertilizers can also be utilized to alleviate the harmful effects of salt stress. This characteristic is particularly crucial for cultivating plants in high-salinity lands. In a recent study, Hashem et al. (2019) attempted to investigate the potential of three seaweeds (*Ulva lactuca Linnaeus*, *Cystoseira* spp., and *Gelidium crinale*) in mitigating the adverse impacts of salt stress (75–150 mM sodium chloride) on canola plants. According to the authors, the application of these seaweeds as soil amendments proved successful in alleviating salinity-induced detrimental effects on canola plants (Hashem et al. 2019). In addition, Hussein et al. (2021) also reported the extracts prepared from seaweeds (*U. fasciata*, *C. compressa*, and *Laurencia obtuse*) have the potential to act as salinity stress alleviant for *Zea mays* and *Vigna sinensis* (Hussein et al. 2021). The potential of *Padina gymnospora* extract to improve the growth rate and productivity of tomato plants under salinity stress was also reported by Hernández-Herrera et al. (2022) in a recent study. Mitigation of salt stress in *Triticum durum* L. (durum wheat) seedlings through the application of liquid seaweed extracts of *Fucus spiralis* were also reported by Latique et al. (2017). According to the authors, the activity of antioxidant enzymes showed a rise corresponding to the increased concentration of algal extract, reaching up to 50%. Consequently, the application of algal treatment has been demonstrated as an efficient method to enhance the growth of wheat seedlings in the presence of salt-stress conditions (Latique et al. 2017). Taken together, these studies provide valuable details about the novel potential of seaweeds for agriculture in terms of improving salt tolerance of major agricultural crops.

7.2.3 Aquaponics and Hydroponics Applications

Marine plant resources can be integrated into nutrient solutions for aquaponics and hydroponics systems. The bioactive compounds in seaweed extracts can provide essential nutrients directly to plants in soilless growing systems. In a recent study, Frassine et al. (2023) attempted to evaluate the growth performance of *Lactuca sativa* L. plants using *Spirulina* waste biomass. According to the authors, spirulina extract-treated lettuce plants had a positive effect on growth and the speed of the plant (Frassine et al. 2023). In another investigation, Jung and Kim (2020) documented the comprehensive reutilization of mixed mackerel and *Undaria pinnatifida* wastewater, highlighting its potential as a high-quality biofertilizer in open-flow lettuce hydroponics (Jung and Kim 2020). The authors further highlighted that the biofertilizer contributes to an increased growth rate and antioxidant activity in lettuce. Additionally, antioxidants present in the biofertilizer accumulate in the lettuce leaves, and the study observed that pathogens did not penetrate into the biofertilizer during open-flow hydroponics.

Collectively, the progress in sustainable and cleaner production from wastewater without causing secondary pollution is still in the developmental stages.

Consequently, the outcomes of the aforementioned studies hold promise not only for wastewater treatment but also for minimizing reliance on chemical fertilizers

7.2.4   STIMULATE ROOT GROWTH AND NUTRIENT UPTAKE

Some marine plant extracts possess chelating properties, helping to make essential trace elements more available to plants. This enhances the efficiency of nutrient uptake and utilization by crops (Rathore et al. 2009). Furthermore, some marine-derived compounds promote root development and branching. Including these compounds in fertilizer formulations can lead to a stronger and more extensive root system, enhancing nutrient absorption and overall plant vigor (Yuan et al. 2019).

7.2.5   BIODEGRADABLE MULCHING MATERIALS

Certain marine plant materials can be processed into biodegradable mulches that, when used in combination with fertilizers, contribute to weed control, soil moisture retention, and improved nutrient availability. In a recent study, Merino et al. (2021) attempted to evaluate the growth parameters of tomato plants by using a mulch suspension prepared from the combination of sodium alginate and three different concentrations of *U. pinnatifida*. (Merino et al. 2021). According to the authors, tomato plants treated with seaweed mulch had a high root mass compared to the nontreated plants. In addition, Zhao et al. (2017) reported the potential of newly developed kelp mulching films to replace traditional plastic mulching films in agriculture (Zhao et al. 2017).

7.2.6   DISEASE RESISTANCE AND PLANT IMMUNITY

Besides the abiotic factors, microorganisms like bacteria, fungi, and viruses impair plant growth and reproduction. Different plant species adopted different defense mechanisms to counteract these living stresses triggered by plant pathogens. The plant's inherent immune system comprises a multifaceted defense mechanism, encompassing pathogen-triggered immunity and effector-triggered immunity against various pathogenic organisms (Jones and Dangl 2006). Pathogen-triggered immunity and effector-triggered immunity prompt localized immune responses, activating either induced systemic resistance or systemic acquired resistance. This defense response involves a complex network of signaling molecules such as ethylene, jasmonic acid, and salicylic acid (Agarwal, Dangariya and Agarwal 2021).

Marine plant extracts are also found to contain compounds that contribute to plant disease resistance and immunity (Table 7.1) (Mukherjee and Patel 2019; Shukla et al. 2021). Incorporating marine extracts into plant fertilizers can enhance the plant's ability to fend off pathogens, reducing the need for chemical interventions (Shukla et al. 2021). For instance, a number of studies report the potential of different marine extracts to inhibit pathogen growth. In a previous study, Jayaraj et al. (2008) reported the potential of *Ascophyllum nodosum* to develop as an anti-fungal agents against *Alternaria radicina* and *Botrytis cinerea* grows in carrot foliage. In addition, Ali, Ramsubhag, and Jayaraman (2022) recently highlighted the potential of *Acanthophora spicifera*, *Codium taylorii*, *Caulerpa serrulata*, *Gracilaria ornata*, and *Sargassum vulgare* to inhibit *Xanthomonas campestris* and *Alternaria solani* in sweet paper.

TABLE 7.1

Applications of Marine Plant Resources in New Plant Fertilizer Products

	Marine Organism	Activity/Application	Reference
1	*Kappaphycus alvarezii* *Sargasum vulgare*	Plant growth stimulation	Melo et al. (2020)
2	*Gelidium crinale*	*Brassica napus L.* growth stimulation	Hashem et al. (2019)
3	Seaweed extracts	*Enhance the growth of Artemisia annua L.*	Ghatas et al. (2021)
4	*Laminaria* and *Ascophyllum nodosum* spp.	*Zea mays growth stimulation*	Ertani et al. (2018)
5	*Ecklonia maxima*	*Biostimulates growth of baby lettuce*	Di Mola et al. (2019)
6	*Ulva lactuca Linnaeus,* *Cystoseira* spp., and *Gelidium crinale*	Salt tolerance activity of *Brassica napus* L.	Hashem et al. (2019)
7	*Ulva fasciata, Cystoseira compressa,* and *Laurencia obtuse*	Salinity stress alleviators for *Vigna sinensis* and *Zea mays*	Hussein et al. (2021)
8	*Padina gymnospora*	Improves the growth and productivity of tomato plants under salinity stress	Hernández-Herrera et al. (2022)
9	*Fucus spiralis*	Alleviation of salt stress in durum wheat (*Triticum durum* L.)	Latique et al. (2017)
10	*Spirulina*	Growth stimulation of *Lactuca sativa*	Frassine et al. (2023)
11	*Undaria pinnatifida*	Growth stimulation of *lettuce*	Jung and Kim (2020)
12	*Kappaphycus alvarezii*	Growth stimulation *Glycine max*	Rathore et al. (2009C)
13	*Saccharina japonica*	Root growth promotion of *Agriophyllum squarrosum* and *Achnatherum splendens*	Yuan et al. (2019)
14	*Undaria pinnatifida*	As a mulch; stimulate the growth of tomato root mass	Merino et al. (2021)
15	Kelp seaweed	As an alternative to plastic mulch	Zhao et al. (2017)
16	*Sargassum vulgare*	Inhibits *Alternaria solani* and *Xanthomonas campestris* pv. *vesicatoria* in sweet paper	Ali, Ramsubhag, and Jayaraman (2022)
17	*Ascophyllum nodosum*	Inhibits the growth of *Alternaria radicina* and *Botrytis cinerea* grows in carrot foliage	Agarwal, Dangariya, and Agarwal (2021)
18	*Sargassum tenerrimum*	*Macrophomina phaseolina* resistance in tomato	Khedia et al. (2020)

7.3 ROLES OF BIOACTIVE MARINE COMPOUNDS IN THE FUNCTIONAL FOOD INDUSTRY

In recent times, there has been considerable interest in extracting and identifying bioactive compounds from marine bioresources for use as nutraceuticals and functional foods (Herath et al. 2018; Sanjeewa and Herath 2023). The significance of these bioactive components as functional ingredients in the food industry has been widely acknowledged (Herath et al. 2018). Among the potential bioresources, edible seaweeds, commonly known as sea vegetables, represent invaluable sources of structurally diverse bioactive compounds with potential applications in the nutraceutical industry. Studies indicate that bioactive compounds from seaweed exhibit various health benefits, including antimicrobial, anti-inflammatory, antioxidant, anticancer, antidiabetic, anti-obesity, anticoagulant, hepatoprotective, antihypertensive, radioprotective, and anti-Alzheimer's, (Kim et al. 2023; Sanjeewa et al. 2023). This section specifically explores bioactive components derived from marine organisms, such as sulfated polysaccharides, fucosterol, lectins, carotenoid pigments, proteins, mycosporine-like amino acids, phlorotannins, and peptides, highlighting their potential health properties and applications in relation to the functional food applications.

7.3.1 ANTIOXIDANT PROPERTIES REPORTED FROM MARINE ORGANISMS

In general, oxidative stress is characterized by a disparity between heightened levels of reactive oxygen species (ROS) and reduced activity of antioxidant systems. Increased oxidative stress can induce damage to the cellular structure and further affect the functions of tissues and organs. However, ROS are required for proper cell activity such as mitochondrial functions (Preiser 2012). Increased oxidative stress has been linked to physiological states, including aging and exercise, as well as a number of pathological illnesses like neurodegenerative diseases, cancer, diabetes inflammatory diseases, cardiovascular disease, and intoxications (Barnham, Masters and Bush 2004; Vaziri and Rodriguez-Iturbe 2006). Marine organisms have been recognized for their notable antioxidant properties, contributing to their potential use in various health-related applications. These antioxidants play a crucial role in neutralizing harmful free radicals, thereby protecting cells from oxidative stress and potential damage. Several edible marine organisms were found to demonstrate significant antioxidant properties.

According to earlier studies, the biological efficacy of unique and structurally diverse compounds produced by marine organisms such as microalgae, seaweeds, sea cucumbers, clams, sea urchins, etc. were found to surpass that of substances derived from terrestrial sources (Perera et al. 2023; Shahidi and Ambigaipalan 2015). Exposed to environmental stresses and fluctuations, marine biota develop defensive mechanisms, giving rise to secondary metabolites and macromolecules recognized for their antioxidant activity. The exploration of marine antioxidants and their potential applications across various industries, including pharmaceuticals, nutraceuticals, cosmetics, and others, has garnered substantial research attention (Freitas et al. 2012). Many antioxidant compounds, such as peptides, polyphenols, polysaccharides, pigments, lipids, etc., exhibit not only antioxidant properties but also additional biological functions, including antimicrobial, wound healing, anticancer, antidiabetic, neuroprotective, anti-fibrotic, anti-Alzheimer, lipid-lowering, sleep-enhancing, and skin protection benefits (Freitas et al. 2012; Galasso, Corinaldesi and Sansone 2017).

7.4 MEDICINAL VALUE OF MARINE BIORESOURCES

Marine bioresources, which encompass various organisms and compounds derived from the ocean, have shown significant medicinal value. The diverse marine environment is a rich source of unique and bioactive compounds that have demonstrated potential in pharmaceutical and medical applications. The following sections are devoted to discussing some of the major biomedical applications of marine-derived compounds.

7.4.1 ANTIBIOTICS

Antibiotics are compounds that are able to act against bacteria and, thus, are used to control or prevent bacterial infections-related diseases and complications (Hall et al. 2023; Nweze et al. 2020). Marine organisms like bacteria, cyanobacteria, fungi, sponges, seaweeds, cnidaria, bryozoa, mollusca, annelida, tunicate, and echinodermata have yielded a number of compounds with antibiotic properties (López, Cepas and Soto 2018; Nweze et al. 2020; Stonik, Makarieva and Shubina 2020). Specifically, sessile or slow-moving marine organisms have developed adaptive defense mechanisms to protect themselves against pathogenic microorganisms. Furthermore, certain marine creatures were discovered to retain symbiotic connections with microbes, bacterial symbionts responsible for the synthesis of antimicrobial compounds (Alves et al. 2020). Despite the vast diversity of marine ecosystems and the promising bioactivities reported, only a few antibiotics identified from marine biota are currently available on the market. There is a need for further exploration in the realm of studies related to marine bacteria if we aim to discover novel antibiotics from marine microorganisms like microalgae. The combination of microbiology and chemistry will be a crucial approach to making advancements in the successful quest for new marine drugs.

7.4.2 ANTIVIRAL PROPERTIES OF MARINE ORGANISMS

A number of studies have demonstrated that marine organisms (seaweeds, fungi, bacteria, sponges, microalgae, green mussels, tunicates) produce a range of metabolites, derived from primary or secondary metabolism, with the promising antiviral properties (Geahchan, Ehrlich and Rahman 2021; Ogodo, Egbuna and Okoronkwo 2022). Moreover, some compounds identified from marine organisms are found to be active against many pathogenic viruses, including Sars-CoV-2 (Geahchan, Ehrlich and Rahman 2021; Riccio et al. 2020). According to previous studies, different classes of compounds isolated from marine organisms such as carbohydrates, polyketides, peptides, lipids, polyphenols, exopolysaccharides, alkaloids, quinones, steroids, terpenoids, polyketones, and zoanthoxanthins have been found to be effective against a number of pathogenic virus strains (Teng et al. 2020).

7.4.3 ANTICANCER COMPOUNDS

The establishment of cancer registries worldwide has prompted a quest for new drugs with the ability to inhibit cancer cells without affecting the normal cells. According to previous studies, conventional anticancer drugs often demonstrated considerable toxicity, affecting both cancer and healthy cells in the affected body part (Lichota and Gwozdzinski 2018). Thus, natural anticancer drugs with low side effects have more demand than synthetic drugs (Sanjeewa et al. 2017). Many compounds and crude extracts separated from marine organisms, particularly certain types of algae, sponges, and soft corals, were found to demonstrate potential anticancer properties. Mostly these isolated compounds either inhibit the growth of carcinoma cells or induce apoptosis (programmed cell death) (Ashkenazi 2008).

Interestingly, according to the epidemiological investigations reported from numerous Asian countries where fish and seafood consumption is high, there is a lower incidence of specific cancers like lung, breast, colorectal, and prostate cancers. These findings have also prompted thorough research into the potential advantages of substances found in consumable marine organisms such as marine invertebrates (mollusks and echinoderms), fin fish species, microorganisms, and seaweeds as agents for preventing cancer (Correia-da-Silva et al. 2017). Remarkably, most of these marine organisms are not only consumed as seafood but are also used as ingredients in the traditional folk medicine of certain East and Southeast Asian regions.

7.4.4   ANTI-INFLAMMATORY AGENTS

Inflammation is part of the body's defense mechanism. *"Inflammation is the process by which the immune system recognizes and removes harmful and foreign stimuli and begins the healing process. Inflammation can be either acute or chronic"* (Gusev and Zhuravleva 2022; Michels da Silva, Langer and Graf 2019). Moreover, heat, swelling, redness, discomfort, and loss of tissue function are some common symptoms of inflammation (Sanjeewa et al. 2021). It has been reported that chronic inflammation plays a role in a number of diseases, such as arthritis, asthma, atherosclerosis, autoimmune disorders, diabetes, and age-related conditions (Coussens and Werb 2002; Germolec et al. 2018). According to previous reports, up to 20% of human cancers are related to chronic, unresolved inflammation caused by irritants, bacterial and viral infections, and autoimmune diseases. Therefore, control of excessive inflammatory reactions is important for sustaining a well-balanced immune system and human health (Sanjeewa et al. 2021).

So far, a number of bioactive metabolites have been isolated/identified from marine organisms such as marine microorganisms (fungi, microalgae, bacteria, and fungi) and macroorganisms (seaweeds, sponges, and coral) to treat inflammatory disorders (Li et al. 2021). The anti-inflammatory marine-based molecules list includes sulfated polysaccharides, proteins, fatty acids, terpenes, etc. Hence, the consumption of these edible marine-based organisms in a regular manner might have the possibility to protect our tissues and organs from many chronic inflammatory diseases (Fernando, Nah and Jeon 2016).

7.4.5   CARDIOVASCULAR MEDICATIONS

Cardiovascular diseases are one of the leading causes of mortality globally. Cardiovascular diseases encompass a wide range of disorders, including those affecting the cardiac muscle and the vascular system that supplies the heart, brain, and other essential organs (Gaziano et al. 2006; Nabel 2003). Furthermore, high blood pressure, high cholesterol, excessive tobacco and alcohol intake, and a lack of vegetables and fruits are well-known risk factors for cardiovascular diseases.

Recently, several studies highlighted that bioactive compounds (polysaccharides, amino acids, proteins, ω-3 polyunsaturated fatty acids, fatty acids, carotenoids, chlorophylls, and peptides) isolated from marine organisms such as seaweeds, microalgae, sponges, oyster, fin fish species, sea cucumbers, etc. (Guo et al. 2015; Lordan, Ross and Stanton 2011; Murray et al. 2018; Ngo et al. 2012). Consumption of these marine organisms may help to reduce the risk of heart disease by reducing risk factors for cardiovascular diseases.

7.5 COSMECEUTICALS FROM MARINE ORGANISMS

The term cosmeceuticals was coined by Albert Kligman to define cosmetic products with biologically active substances that carry either medicinal or drug-like benefits (Sanjeewa et al. 2016). Moreover, cosmetic products are commonly used to enhance skin appearance and maintain skin care. Recently, the demand for cosmeceuticals with natural ingredients has increased compared to cosmeceuticals with synthetic active ingredients due to the low side effects associated with natural ingredients (Eom and Kim 2013).

Marine natural products offer a diverse range of chemicals that can be utilized in the design and development of potentially beneficial therapeutic products. Furthermore, numerous marine compounds exhibit promising properties as cosmeceutical agents due to their antiaging, antioxidant, anti-wrinkling, antiallergic, anti-inflammatory, tyrosinase inhibitory, MMP inhibitory, and ultraviolet protective activities (Agrawal et al. 2018; Sanjeewa et al. 2016; Schneider et al. 2020; Thomas and Kim 2013).

Polyphenols, carotenoids, exopolysaccharides, benzodiazepine alkaloids, mycosporine, and mycosporine-like amino acids isolated from marine macroorganisms and microorganisms have demonstrated potential bioactive properties. These include antioxidant, anti-inflammatory, antiaging, antiallergic, tyrosinase inhibitory, anti-wrinkling, MMP inhibitory, and ultraviolet protective activities, making them promising candidates for development as active ingredients in cosmeceutical formulas (Corinaldesi et al. 2017; Sanjeewa et al. 2016). It is widely acknowledged that these biological effects are closely linked to their cosmeceutical applications.

7.6 APPLICATIONS OF GELATIN SEPARATED FROM MARINE FISH BY-PRODUCTS

Gelatin is a natural polymer produced as a result of hydrolytic degradation of protein from collagen (Kumosa, Zetterberg and Schouenborg 2018). The chemical composition of gelatin comprises various polypeptide chains, including α-chains (single chains), β-chains (two α-chains covalently linked), and γ-chains (three covalently linked α-chains), each having respective molar masses of approximately 90×10^3, 180×10^3, and 300×10^3 g/mol. During processing, heating causes gelatin to dissolve into colloids. However, at temperatures below 40 °C, gelatin transforms into a solid state. If an aqueous gelatin solution is boiled extensively, its properties undergo alteration due to the degradation of structure and will lose the gelatinous properties at lower temperatures (Alipal et al. 2021).

Gelatin serves as a multipurpose component utilized in the food, cosmetic, pharmaceutical, and photographic sectors. Gelatin functions as a gelling agent, stabilizer, microencapsulating agent, emulsifier, thickener, and film-former. While the primary sources of gelatin are the skins of pigs, cattle bones, and bovine hides, these mammalian-derived gelatins face rejection by some consumers due to sociocultural, religious, or health-related concerns (Karim and Bhat 2009, 2008; Lin et al. 2017). As an alternative to the aforementioned gelatin sources, gelatins extracted from the by-products of aquatic food processing of cold water fish (cod, Alaska pollock, and salmon) and warm water (catfish, tuna, Nile perch, tilapia, and shark) fish skins, bones, and fins. Moreover, gelatin extracted from fish by-products has gained attention across the gelatin-utilizing industries (Karim and Bhat 2009; Kim and Mendis 2006; Lin et al. 2017). This section highlights the wide-ranging applications of gelatin derived from marine fish by-products, emphasizing its diverse utility in various industries.

7.6.1 General Extraction Protocol of Fish Gelatin

Most gelatin production procedures are divided into three stages: raw material pretreatment, gelatin extraction, and purification and drying (Karim and Bhat 2009). Gelatin is often extracted from fish by-products at temperatures ranging from 55°C to 70°C. Collagen can be converted into gelatin through acid and alkali treatments. The acid process is predominantly used for extracting gelatin from fish skins, where collagen undergoes hydrolysis at pH 4, resulting in what is referred to as type A gelatin. Conversely, the alkali method is applied to extract gelatin from bovine hide, known as type B gelatin. After the extraction, the gelatin solution is filtered to remove insoluble materials such as unhydrolyzed collagen and lipids. The subsequent purification stage employs ion exchange or ultrafiltration columns to eliminate inorganic salts. The final steps involve dehydration, sterilization, and drying (Usman et al. 2021).

7.6.2 Food Industry Applications of Gelatin

Gelatin stands out as a distinctive and special hydrocolloid among those utilized in the food industry, showcasing versatility by performing numerous functions and finding a broad spectrum of applications across various industries (Karim and Bhat 2008). Specifically, gelatin is able to improve the texture, viscosity, and stability of liquid- and semisolid-type foods. Moreover, gelatin is used to form soft, chewy candies, and gelatin has also found new use as an emulsifier and extender in the manufacturing of low-fat margarine products (Keenan 2012). However, food-grade gelatin replacement has been a major issue in the last few decades due to various sociocultural reasons and the onset of zoonotic disease in cattle (Nitsuwat et al. 2021).

When exploring potential alternative sources, fish by-products emerge as promising candidates. In particular, the substantial by-products generated from fish bones and skins during fish processing, where product yields typically range from 30% to

50%, offer a significant opportunity to extract food-grade gelatin (Nitsuwat et al. 2021). Besides the potentials at present, fish gelatin is rarely used in the food industry compared to gelatin of bovine and porcine origins.

Compared to bovine or other mammalian gelatin, fish gelatin shows weaker gelation and rheological properties, which causes reduction in the practical application (Alfaro et al. 2014). In addition, fish gelatin is reported to show poor gelling and melting temperatures, storage moduli, and weaker gel strengths compared to mammalian gelatin gels (Derkach et al. 2021). In addition, the fishy smell also reduces the food applications of fish gelatin. Thus, further study is needed to better understand the specific mechanism of such alterations and broaden the applications of fish gelatine in the food industry.

7.6.3 OTHER POTENTIAL APPLICATIONS OF FISH GELATIN

Besides the food application, fish gelatin was found to be useful in a number of fields. According to recent studies, fish gelatin can be used in cosmetic applications (Nurilmala et al. 2022). Gelatin isolated from fish by-products has been widely investigated for cosmeceutical applications due to the biological properties reported from fish gelatin such as antimicrobial, antioxidant, matrix metalloproteinase inhibitory properties, UV-protective, and anti-photoaging activities (Rajabimashhadi et al. 2023; Viji et al. 2019). These biological activities of the marine fish gelatin have led to the development of functional cosmetics for different cosmeceutical purposes such as antiaging, skincare, and anti-wrinkle (Venkatesan et al. 2017).

Biomedical applications of fish gelatin such as wound healing and dressing, tissue engineering, gene therapy, implants, and bone substitutes have also been reported by different research studies. Moreover, fish gelatin holds promise for pharmaceutical applications, encompassing the production of capsules, microparticle/oil coating, tablet coating, emulsion stabilization, and various drug delivery applications like microspheres, nanospheres, microneedles, scaffolds, and hydrogels. Key findings indicate that fish gelatin is immunologically safe, guards against potential transmission of diseases like bovine spongiform encephalopathy and foot and mouth diseases, and offers both economic and environmental advantages.

7.7   FUTURE DIRECTIONS

Marine organisms and their secondary metabolites have shown promise for various applications across industries. Specifically, the bioactive metabolites present in marine organisms hold significant potential for development into functional foods, cosmeceuticals, nutraceuticals, and pharmaceuticals due to their demonstrated bioactivities. Additionally, marine bioresources can serve as a major ingredient in animal feed formulations and find applications in agriculture, such as in biofertilizers and pesticides. However, the full utilization of marine bioresources for most applications faces challenges, primarily stemming from issues in the supply chain, compliance with laws and regulations, high processing costs, seasonal dependency of bioactive ingredients, and sociocultural beliefs. Addressing these challenges is crucial for the successful commercialization and scaling up of production to meet industry requirements.

REFERENCES

AbdElgawad, H., G. Zinta, M. M. Hegab, R. Pandey, H. Asard, and W. Abuelsoud. 2016. "High salinity induces different oxidative stress and antioxidant responses in maize seedlings organs." *Frontiers in Plant Science* 7:276. doi: 10.3389/fpls.2016.00276

Agarwal, P. K., M. Dangariya, and P. Agarwal. 2021. "Seaweed extracts: Potential biodegradable, environmentally friendly resources for regulating plant defence." *Algal Research* 58:102363. doi: 10.1016/j.algal.2021.102363

Agrawal, S., A. Adholeya, C. J. Barrow, and S. K. Deshmukh. 2018. "Marine fungi: An untapped bioresource for future cosmeceuticals." *Phytochemistry Letters* 23:15–20. doi: 10.1016/j.phytol.2017.11.003

Alipal, J., N. A. S. Mohd Pu'ad, T. C. Lee, N. H. M. Nayan, N. Sahari, H. Basri, M. I. Idris, and H. Z. Abdullah. 2021. "A review of gelatin: Properties, sources, process, applications, and commercialisation." *Materials Today: Proceedings* 42:240–250. doi: 10.1016/j.matpr.2020.12.922

Ali, O., A. Ramsubhag, and J. Jayaraman. 2022. "Application of extracts from Caribbean seaweeds improves plant growth and yields and increases disease resistance in tomato and sweet pepper plants." *Phytoparasitica* 51 (4):727–745. doi: 10.1007/s12600-022-01035-w

Alves, E., M. Dias, D. Lopes, A. Almeida, M. D. Domingues, and F. Rey. 2020. "Antimicrobial lipids from plants and marine organisms: An overview of the current state-of-the-art and future prospects." *Antibiotics* 9 (8):441. doi: 10.3390/antibiotics9080441

Ashkenazi, A. 2008. "Targeting the extrinsic apoptosis pathway in cancer." *Cytokine Growth Factor Reviews* 19 (3-4):325–31. doi: 10.1016/j.cytogfr.2008.04.001

Barnham, K. J., C. L. Masters, and A. I. Bush. 2004. "Neurodegenerative diseases and oxidative stress." *Nature Reviews Drug Discovery* 3 (3):205–214. doi: 10.1038/nrd1330

Chanda, M. J., N. Merghoub, and H. El Aroussi. 2019. "Microalgae polysaccharides: The new sustainable bioactive products for the development of plant bio-stimulants?" *World Journal of Microbiology and Biotechnology* 35 (11):177. doi: 10.1007/s11274-019-2745-3

Chiaiese, P., G. Corrado, G. Colla, M. C. Kyriacou, and Y. Rouphael. 2018. "Renewable sources of plant biostimulation: Microalgae as a sustainable means to improve crop performance." *Frontiers in Plant Science* 9:1782.

Cook, J., J. Zhang, J. Norrie, B. Blal, and Z. Cheng. 2018. "Seaweed extract (Stella Maris®) activates innate immune responses in *Arabidopsis thaliana* and protects host against bacterial pathogens." *Marine Drugs* 16 (7):221. doi: 10.3390/md16070221

Corinaldesi, C., G. Barone, F. Marcellini, A. Dell'Anno, and R. Danovaro. 2017. "Marine microbial-derived molecules and their potential use in cosmeceutical and cosmetic products." *Marine Drugs* 15 (4):118. doi: 10.3390/md15040118

Correia-da-Silva, M., E. Sousa, M. M. M. Pinto, and A. Kijjoa. 2017. "Anticancer and cancer preventive compounds from edible marine organisms." *Seminars in Cancer Biology* 46:55–64. doi: 10.1016/j.semcancer.2017.03.011

Coussens, L. M., and Z. Werb. 2002. "Inflammation and cancer." *Nature* 420 (6917):860–867. doi: 10.1038/nature01322

Daniotti, S., and I. Re. 2021. "Marine biotechnology: Challenges and development market trends for the enhancement of biotic resources in industrial pharmaceutical and food applications. A statistical analysis of scientific literature and business models." *Marine Drugs* 19 (2):61. doi: 10.3390/md19020061

da Trindade Alfaro, A., E. Balbinot, C. I. Weber, I. B. Tonial, and A. Machado-Lunkes. 2014. "Fish gelatin: Characteristics, functional properties, applications and future potentials." *Food Engineering Reviews* 7 (1):33–44. doi: 10.1007/s12393-014-9096-5

Derkach, S. R., D. S. Kolotova, N. G. Voron'ko, E. D. Obluchinskaya, and A. Y. Malkin. 2021. "Rheological properties of fish gelatin modified with sodium alginate." *Polymers* 13 (5):743. doi: 10.3390/polym13050743

Di Mola, I., E. Cozzolino, L. Ottaiano, M. Giordano, Y. Rouphael, G. Colla, and M. Mori. 2019. "Effect of vegetal- and seaweed extract-based biostimulants on agronomical and leaf quality traits of plastic tunnel-grown baby lettuce under four regimes of nitrogen fertilization." *Agronomy* 9 (10):571. doi: 10.3390/agronomy9100571

Eom, S.-H., and S.-K. Kim. 2013. "Cosmeceutical Applications from Marine Organisms." In Cosmeceuticals and Cosmetic Practice, 200–208.

Ertani, A., O. Francioso, A. Tinti, M. Schiavon, D. Pizzeghello, and S. Nardi. 2018. "Evaluation of seaweed extracts from *Laminaria and Ascophyllum nodosum* spp. as biostimulants in *Zea mays* L. using a combination of chemical, biochemical and morphological approaches." *Frontiers in Plant Science* 9:428.

Fernando, I. P. S., J. W. Nah, and Y. J. Jeon. 2016. "Potential anti-inflammatory natural products from marine algae." *Environmental Toxicology and Pharmacology* 48:22–30. doi: 10.1016/j.etap.2016.09.023

Frassine, D., R. Braglia, F. Scuderi, E. L. Redi, A. Gismondi, G. Di Marco, L. Rugnini, and A. Canini. 2023. "Sustainability in aquaponics: Industrial spirulina waste as a biofertilizer for *Lactuca sativa* L. plants." *Plants* 12 (23). doi: 10.3390/plants12234030

Freitas, A. C., D. Rodrigues, T. A. Rocha-Santos, A. M. Gomes, and A. C. Duarte. 2012. "Marine biotechnology advances towards applications in new functional foods." *Biotechnology Advances* 30 (6):1506–1515. doi: 10.1016/j.biotechadv.2012.03.006

Galasso, C., C. Corinaldesi, and C. Sansone. 2017. "Carotenoids from marine organisms: Biological functions and industrial applications." *Antioxidants* 6 (4):96. doi: 10.3390/antiox6040096

Gaziano, T., K. S. Reddy, F. Paccaud, S. Horton, and V. Chaturvedi. 2006. "Chapter 33: Cardiovascular Disease." In Disease Control Priorities in Developing Countries, edited by D. T. Jamison. Oxford University Press.

Geahchan, S., H. Ehrlich, and M. A. Rahman. 2021. "The anti-viral applications of marine resources for COVID-19 treatment: An overview." *Marine Drugs* 19 (8):409. doi: 10.3390/md19080409

Germolec, D. R., K. A. Shipkowski, R. P. Frawley, and E. Evans. 2018. "Markers of Inflammation." In *Immunotoxicity Testing: Methods and Protocols*, edited by J. C. DeWitt, C. E. Rockwell, and C. C. Bowman, 57–79. New York, NY: Springer New York.

Ghatas, Y., M. Ali, M. Elsadek, and Y. Mohamed. 2021. "Enhancing growth, productivity and artemisinin content of *Artemisia annua* L. plant using seaweed extract and micronutrients." *Industrial Crops and Products* 161:113202. doi: 10.1016/j.indcrop.2020.113202

Guo, Y., Y. Ding, F. Xu, B. Liu, Z. Kou, W. Xiao, and J. Zhu. 2015. "Systems pharmacology-based drug discovery for marine resources: An example using sea cucumber (Holothurians)." *Journal of Ethnopharmacology* 165:61–72. doi: 10.1016/j.jep.2015.02.029

Gusev, E., and Y. Zhuravleva. 2022. "Inflammation: A new look at an old problem." *International Journal of Molecular Sciences* 23 (9):4596. doi: 10.3390/ijms23094596

Hall, P. P., H. R. Wermuth, and C. Calhoun, G. A. 2023. *Antibiotics*, StatPearls Publishing.

Hashem, H. A., H. A. Mansour, S. A. El-Khawas, and R. A. Hassanein. 2019. "The potentiality of marine macro-algae as bio-fertilizers to improve the productivity and salt stress tolerance of canola (*Brassica napus* L.) plants." *Agronomy* 9 (3):146. doi: 10.3390/agronomy9030146

Herath, K. M., J. H. Lee, J. Cho, A. Kim, S. M. Shin, B. Kim, Y. J. Jeon, and Y. Jee. 2018. "Immunostimulatory effect of pepsin enzymatic extract from *Porphyra yezoensis* on murine splenocytes." *Journal of the Science of Food and Agriculture* 98 (9): 3400–3408. doi: 10.1002/jsfa.8851

Hernández-Herrera, R. M., C. V. Sánchez-Hernández, P. A. Palmeros-Suárez, H. Ocampo-Alvarez, F. Santacruz-Ruvalcaba, I. D. Meza-Canales, and A. Becerril-Espinosa. 2022.

"Seaweed extract improves growth and productivity of tomato plants under salinity stress." *Agronomy* 12 (10). doi: 10.3390/agronomy12102495

Hussein, M. H., E. Eltanahy, A. F. Al Bakry, N. Elsafty, and M. M. Elshamy. 2021. "Seaweed extracts as prospective plant growth bio-stimulant and salinity stress alleviator for *Vigna sinensis* and *Zea mays*." *Journal of Applied Phycology* 33 (2):1273–1291. doi: 10.1007/s10811-020-02330-x

Imhoff, J. F., A. Labes, and J. Wiese. 2011. "Bio-mining the microbial treasures of the ocean: New natural products." *Biotechnology Advances* 29 (5):468–482. doi: 10.1016/j.biotechadv.2011.03.001

Innes, J. K., and P. C. Calder. 2020. "Marine omega-3 (n-3) fatty acids for cardiovascular health: An update for 2020." *International Journal of Molecular Sciences* 21 (4):1362. doi: 10.3390/ijms21041362

Jayaraj, J., A. Wan, M. Rahman, Z. K. Punja. 2008. "Seaweed extract reduces foliar fungal diseases on carrot." *Crop Protection* 27 (10):1360–1366. doi: 10.1016/j.cropro.2008.05.005

Jones, J. D., and J. L. Dangl. 2006. "The plant immune system." *Nature* 444 (7117):323–329. doi: 10.1038/nature05286

Jung, H. Y., and J. K. Kim. 2020. "Complete reutilisation of mixed mackerel and brown seaweed wastewater as a high-quality biofertiliser in open-flow lettuce hydroponics." *Journal of Cleaner Production* 247:119081. doi: 10.1016/j.jclepro.2019.119081

Karim, A. A., and R. Bhat. 2008. "Gelatin alternatives for the food industry: Recent developments, challenges and prospects." *Trends in Food Science & Technology* 19 (12): 644–656. doi: 10.1016/j.tifs.2008.08.001

Karim, A. A., and R. Bhat. 2009. "Fish gelatin: Properties, challenges, and prospects as an alternative to mammalian gelatins." *Food Hydrocolloids* 23 (3):563–576. doi: 10.1016/j.foodhyd.2008.07.002

Keenan, T. R. 2012. "Gelatin." In *Polymer Science: A Comprehensive Reference*, edited by Krzysztof Matyjaszewski and Martin Möller, 237–247. Amsterdam: Elsevier.

Khedia, J., M. Dangariya, A. K. Nakum, P. Agarwal, A. Panda, A. K. Parida, D. R. Gangapur, R. Meena, and P. K. Agarwal. 2020. "Sargassum seaweed extract enhances *Macrophomina phaseolina* resistance in tomato by regulating phytohormones and antioxidative activity." *Journal of Applied Phycology* 32 (6):4373–4384. doi: 10.1007/s10811-020-02263-5

Kim, S.-K., and E. Mendis. 2006. "Bioactive compounds from marine processing byproducts: A review." *Food Research International* 39 (4):383–393. doi: 10.1016/j.foodres.2005.10.010

Kim, H. J., J. Yang, K. Herath, Y. J. Jeon, Y. O. Son, D. Kwon, H. J. Kim, and Y. Jee. 2023. "Oral administration of *Sargassum horneri* suppresses particulate matter-induced oxidative DNA damage in alveolar macrophages of allergic airway inflammation: Relevance to PM-mediated M1/M2 AM polarization." *Molecular Nutrition & Food Research* 67 (24):e2300462. doi: 10.1002/mnfr.202300462

Kumosa, L. S., V. Zetterberg, and J. Schouenborg. 2018. "Gelatin promotes rapid restoration of the blood brain barrier after acute brain injury." *Acta Biomaterialia* 65:137–149. doi: 10.1016/j.actbio.2017.10.020

Latique, S., E. M. Aymen, C. Halima, H. Chérif, and E. K. Mimoun. 2017. "Alleviation of salt stress in durum wheat (*Triticum durum* L.) seedlings through the application of liquid seaweed extracts of *Fucus spiralis*." *Communications in Soil Science and Plant Analysis* 48 (21):2582–2593. doi: 10.1080/00103624.2017.1416136

Lichota, A., and K. Gwozdzinski. 2018. "Anticancer activity of natural compounds from plant and marine environment." *International Journal of Molecular Sciences* 19 (11):3533. doi: 10.3390/ijms19113533

Li, C.-Q., Q.-Y. Ma, X.-Z. Gao, X. Wang, and B.-L. Zhang. 2021. "Research progress in anti-inflammatory bioactive substances derived from marine microorganisms, sponges, algae, and corals." *Marine Drugs* 19 (10). doi: 10.3390/md19100572

Lin, L., J. M. Regenstein, S. Lv, J. Lu, and S. Jiang. 2017. "An overview of gelatin derived from aquatic animals: Properties and modification." *Trends in Food Science & Technology* 68:102–112. doi: 10.1016/j.tifs.2017.08.012

López, Y., V. Cepas, and S. M. Soto. 2018. "The Marine Ecosystem as a Source of Antibiotics." In *Grand Challenges in Marine Biotechnology*, edited by P. H. Rampelotto and A. Trincone, 3–48. Cham: Springer International Publishing.

Lordan, S., R. P. Ross, and C. Stanton. 2011. "Marine bioactives as functional food ingredients: Potential to reduce the incidence of chronic diseases." *Marine Drugs* 9 (6):1056–1100. doi: 10.3390/md9061056

Melo, P., C. Abreu, K. Bahcevandziev, G. Araujo, and L. Pereira. 2020. "Biostimulant effect of marine macroalgae bioextract on pepper grown in greenhouse." *Applied Sciences* 10 (11):4052. doi: 10.3390/app10114052

Merino, D., M. F. Salcedo, A. Y. Mansilla, C. A. Casalongué, and V. A. Alvarez. 2021. "Development of sprayable sodium alginate-seaweed agricultural mulches with nutritional benefits for substrates and plants." *Waste and Biomass Valorization* 12 (11):6035–6043. doi: 10.1007/s12649-021-01441-x

Michels da Silva, D., H. Langer, and T. Graf. 2019. "Inflammatory and molecular pathways in heart failure-ischemia, HFpEF and transthyretin cardiac amyloidosis." *International Journal of Molecular Sciences* 20 (9):2322. doi: 10.3390/ijms20092322

Mukherjee, A., and J. S. Patel. 2019. "Seaweed extract: Biostimulator of plant defense and plant productivity." *International Journal of Environmental Science and Technology* 17 (1):553–558. doi: 10.1007/s13762-019-02442-z

Murray, M., A. L. Dordevic, L. Ryan, and M. P. Bonham. 2018. "An emerging trend in functional foods for the prevention of cardiovascular disease and diabetes: Marine algal polyphenols." *Critical Reviews in Food Science and Nutrition* 58 (8):1342–1358. doi: 10.1080/10408398.2016.1259209

Nabel, E. G. 2003. "Cardiovascular disease." *New England Journal of Medicine* 349 (1):60–72. doi: 10.1056/NEJMra035098

Nabti, E., B. Jha, and A. Hartmann. 2016. "Impact of seaweeds on agricultural crop production as biofertilizer." *International Journal of Environmental Science and Technology* 14 (5):1119–1134. doi: 10.1007/s13762-016-1202-1

Nayaka, S., K. Toppo, and S. Verma. 2017. "Adaptation in Algae to Environmental Stress and Ecological Conditions." In *Plant Adaptation Strategies in Changing Environment*, edited by V. Shukla, S. Kumar, and N. Kumar, 103–115. Singapore: Springer Singapore.

Nedumaran, T. 2017. "Seaweed: A Fertilizer for Sustainable Agriculture." In *Sustainable Agriculture Towards Food Security*, edited by A. Dhanarajan, 159–174. Singapore: Springer.

Ngo, D. H., T. S. Vo, D. N. Ngo, I. Wijesekara, and S. K. Kim. 2012. "Biological activities and potential health benefits of bioactive peptides derived from marine organisms." *International Journal of Biological Macromolecules* 51 (4):378–383. doi: 10.1016/j.ijbiomac.2012.06.001

Nitsuwat, S., P. Zhang, K. Ng, and Z. Fang. 2021. "Fish gelatin as an alternative to mammalian gelatin for food industry: A meta-analysis." *LWT* 141:110899. doi: 10.1016/j.lwt.2021.110899

Nurilmala, M., H. Suryamarevita, H. Husein Hizbullah, A. M. Jacoeb, and Y. Ochiai. 2022. "Fish skin as a biomaterial for halal collagen and gelatin." *Saudi Journal of Biological Sciences* 29 (2):1100–1110. doi: 10.1016/j.sjbs.2021.09.056

Nweze, J. A., F. N. Mbaoji, G. Huang, Y. Li, L. Yang, Y. Zhang, S. Huang, L. Pan, and D. Yang. 2020. "Antibiotics development and the potentials of marine-derived compounds to stem the tide of multidrug-resistant pathogenic bacteria, fungi, and protozoa." *Marine Drugs* 18 (3). doi: 10.3390/md18030145

Ogodo, A. C., C. Egbuna, and C. U. Okoronkwo. 2022. "Marine Organisms as a Potential Source of Antiviral Drugs for the Treatment of Coronavirus Infections and Other

Viral Diseases." In *Coronavirus Drug Discovery*, edited by Chukwuebuka Egbuna, 207–224. Elsevier.

Pereira, L., and J. Cotas. 2019. "Historical Use of Seaweed as an Agricultural Fertilizer in the European Atlantic Area." In *Seaweeds as Plant Fertilizer, Agricultural Biostimulants and Animal Fodder*, 1–22. CRC Press.

Perera, R. M. T. D., K. H. I. N. M. Herath, K. K. A. Sanjeewa, and T. U. Jayawardena. 2023. "Recent reports on bioactive compounds from marine cyanobacteria in relation to human health applications." *Life* 13 (6). doi: 10.3390/life13061411

Preiser, J. C. 2012. "Oxidative stress." *Journal of Parenteral and Enteral Nutrition* 36 (2):147–54. doi: 10.1177/0148607111434963

Raghunandan, B. L., R. V. Vyas, H. K. Patel, and Y. K. Jhala. 2019. "Perspectives of Seaweed as Organic Fertilizer in Agriculture." In *Soil Fertility Management for Sustainable Development*, edited by D. G. Panpatte and Y. K. Jhala, 267–289. Singapore: Springer Singapore.

Rajabimashhadi, Z., N. Gallo, L. Salvatore, and F. Lionetto. 2023. "Collagen derived from fish industry waste: Progresses and challenges." *Polymers* 15 (3):544. doi: 10.3390/polym15030544

Rathore, S. S., D. R. Chaudhary, G. N. Boricha, A. Ghosh, B. P. Bhatt, S. T. Zodape, and J. S. Patolia. 2009. "Effect of seaweed extract on the growth, yield and nutrient uptake of soybean (*Glycine max*) under rainfed conditions." *South African Journal of Botany* 75 (2):351–355. doi: 10.1016/j.sajb.2008.10.009

Riccio, G., N. Ruocco, M. Mutalipassi, M. Costantini, V. Zupo, D. Coppola, D. de Pascale, and C. Lauritano. 2020. "Ten-year research update review: Antiviral activities from marine organisms." *Biomolecules* 10 (7):1007. doi: 10.3390/biom10071007

Sanjeewa, K. K. A., and K. Herath. 2023. "Bioactive secondary metabolites in sea cucumbers and their potential to use in the functional food industry." *Fisheries and Aquatic Sciences* 26 (2):69–86. doi: 10.47853/FAS.2023.e6

Sanjeewa, K. K. A., K. H. I. N. M. Herath, Y.-S. Kim, Y.-J. Jeon, and S.-K. Kim. 2023. "Enzyme-assisted extraction of bioactive compounds from seaweeds and microalgae." *Trends in Analytical Chemistry* 167:117266. doi: 10.1016/j.trac.2023.117266

Sanjeewa, K. K. A., K. H. I. N. M. Herath, H.-W. Yang, C. S. Choi, and Y.-J. Jeon. 2021. "Anti-inflammatory mechanisms of fucoidans to treat inflammatory diseases: A review." *Marine Drugs* 19 (12). doi: 10.3390/md19120678

Sanjeewa, K. K. A., E. A. Kim, K. T. Son, and Y. J. Jeon. 2016. "Bioactive properties and potentials cosmeceutical applications of phlorotannins isolated from brown seaweeds: A review." *Journal of Photochemistry and Photobiology B* 162:100–105. doi: 10.1016/j.jphotobiol.2016.06.027

Sanjeewa, K. K. A., J. S. Lee, W. S. Kim, and Y. J. Jeon. 2017. "The potential of brown-algae polysaccharides for the development of anticancer agents: An update on anticancer effects reported for fucoidan and laminaran." *Carbohydrate Polymers* 177:451–459. doi: 10.1016/j.carbpol.2017.09.005

Schneider, G., F. L. Figueroa, J. Vega, P. Chaves, F. Álvarez-Gómez, N. Korbee, and J. Bonomi-Barufi. 2020. "Photoprotection properties of marine photosynthetic organisms grown in high ultraviolet exposure areas: Cosmeceutical applications." *Algal Research* 49:101956. doi: 10.1016/j.algal.2020.101956

Shahidi, F., and P. Ambigaipalan. 2015. "Novel functional food ingredients from marine sources." *Current Opinion in Food Science* 2:123–129. doi: 10.1016/j.cofs.2014.12.009

Shen, P., Z. Yin, G. Qu, and C. Wang. 2018. "Fucoidan and Its Health Benefits." In *Bioactive Seaweeds for Food Applications*, edited by Y. Qin, 223–238. Academic Press.

Shukla, P. S., T. Borza, A. T. Critchley, and B. Prithiviraj. 2021. "Seaweed-based compounds and products for sustainable protection against plant pathogens." *Marine Drugs* 19 (2):59. doi: 10.3390/md19020059

Sotelo, C. G., M. Blanco, P. Ramos, J. A. Vázquez, and R. I. Perez-Martin. 2021. "Sustainable sources from aquatic organisms for cosmeceuticals ingredients." *Cosmetics* 8 (2):48. doi: 10.3390/cosmetics8020048

Stonik, V. A., T. N. Makarieva, and L. K. Shubina. 2020. "Antibiotics from marine bacteria." *Biochemistry (Mosc)* 85 (11):1362–1373. doi: 10.1134/S0006297920110073

Teng, Y. F., L. Xu, M. Y. Wei, C. Y. Wang, Y. C. Gu, and C. L. Shao. 2020. "Recent progresses in marine microbial-derived antiviral natural products." *Archives of Pharmacal Research* 43 (12):1215–1229. doi: 10.1007/s12272-020-01286-3

Thomas, N. V., and S.-K. Kim. 2013. "Beneficial effects of marine algal compounds in cosmeceuticals." *Marine Drugs* 11 (1):146–164. doi: 10.3390/md11010146

Usman, M., A. Sahar, M. Inam-Ur-Raheem, U. U. Rahman, A. Sameen, and R. M. Aadil. 2021. "Gelatin extraction from fish waste and potential applications in food sector." *International Journal of Food Science & Technology* 57 (1):154–163. doi: 10.1111/ijfs.15286

Vaziri, N. D., and B. Rodriguez-Iturbe. 2006. "Mechanisms of disease: Oxidative stress and inflammation in the pathogenesis of hypertension." *Nature Clinical Practice Nephrology* 2 (10):582–593. doi: 10.1038/ncpneph0283

Venkatesan, J., S. Anil, S.-K. Kim, and M. S. Shim. 2017. "Marine fish proteins and peptides for cosmeceuticals: A review." *Marine Drugs* 15 (5):143. doi: 10.3390/md15050143

Viji, P., T. S. Phannendra, D. Jesmi, B. Madhusudana Rao, P. H. Dhiju Das, and N. George. 2019. "Functional and antioxidant properties of gelatin hydrolysates prepared from skin and scale of sole fish." *Journal of Aquatic Food Product Technology* 28 (10):976–986. doi: 10.1080/10498850.2019.1672845

Yuan, M., H. Xiao, R. Wang, Y. Duan, and Q. Cao. 2019. "Effects of changes in precipitation pattern and of seaweed fertilizer addition on plant traits and biological soil crusts." *Journal of Applied Phycology* 31 (6):3791–3802. doi: 10.1007/s10811-019-01838-1

Zhao, Y., J. Qiu, J. Xu, X. Gao, and X. Fu. 2017. "Effects of crosslinking modes on the film forming properties of kelp mulching films." *Algal Research* 26:74–83. doi: 10.1016/j.algal.2017.07.006

8 Circular Economy Principles Related to the Blue Bioeconomy

8.1 MARINE WASTE MANAGEMENT AND RECYCLING

Marine waste management and recycling constitute pivotal pillars in the multifaceted battle against the escalating environmental threats facing our oceans. Marine debris has been defined as follows: 'Any persistent manufactured or processed solid material discarded, disposed of or abandoned in the marine and coastal environment, has been highlighted as a contaminant of global environmental and economic concern' (Agamuthu et al. 2019). In general plastic, paper, metal, cloth, glass, and rubber are considered major categories of marine waste. This chapter discusses marine plastic waste management, prospects, and future directions in depth.

8.1.1 PLASTIC AS MARINE CONTAMINANT

Plastic is the main contaminant in the ocean environment, and according to Statista, manufacturing of plastics has increased steadily over the previous 75 years, rising from 1.5 million metric tons (Mt) in the 1950s to 390 million Mt in 2021 (Statista 2023). Plastics find extensive applications across diverse sectors, including building construction, transport, packaging, electronics, automotive manufacturing, and agriculture (Hofmann et al. 2023; Pradeep et al. 2017; Roy et al. 2021; Sangroniz et al. 2019). The widespread use of plastics brings about significant societal benefits due to their versatility and applicability in various domains. However, despite their utility, plastics as a commodity have increasingly become a major contaminant of marine ecosystems (Cressey 2016; Okoffo et al. 2021). Recent statistics indicate that around 14 million tons of plastic pile up in the oceans annually, endangering food safety, marine ecology, and economic operations of the whole planet (IUCN 2021).

Plastics enter marine ecosystems through several pathways such as littering, atmospheric transport, and illegal dumping, as well as discharge from storm drains, rivers, and sewage outflows (Figure 8.1) (Santos et al. 2022). Plastic garbage is also occasionally directly abandoned at sea during fishing, shipping, and aquaculture, operations (Lebreton et al. 2017). Due to the persistence, long-lasting nature, and large volume, marine plastic waste is becoming a major threat to the health of marine organisms as well as humans (Phelan et al. 2020). Furthermore, marine mammals frequently ingest marine plastic waste that seems to be food to those organisms. These floating marine water matters such as waste plastics operate as carriers for chemicals and other marine environmental pollutants and are

DOI: 10.1201/9781003477365-8

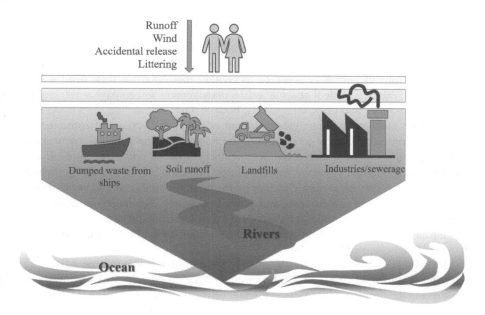

FIGURE 8.1   Major sources of marine plastic pollution.

subsequently released into the marine environment. Additionally, floating plastic debris serves as a potential vector for the introduction of invasive marine microorganisms and macroorganisms into different marine ecosystems (Agamuthu et al. 2019).

Other than the larger sizes, plastic particles size less than 5 mm are considered as microplastic and one of the greatest potential threats to the marine ecosystems (Laskar and Kumar 2019). Plastic waste of various sizes accumulates in the marine environment as a result of a number of degrading processes. Plastic may be destroyed by UV radiation, weathering, current, mechanical–physical, and biodegradation (Lestari and Trihadiningrum 2019). The tiny particles resulting from different plastic degradation processes have the ability to infiltrate tissues and accumulate in organs (Yuan, Nag and Cummins 2022). Their capacity to traverse biological membranes poses a threat to cell functioning, affecting various cell types, including blood cells, and negatively affecting the photosynthesis ability of algae (Wu et al. 2019; Yuan, Nag and Cummins 2022).

Furthermore, some microplastics have been demonstrated to contain reproductive carcinogens, mutagens, and toxins. These harmful matters can accumulate within the food chain at various trophic levels. Research indicates plastic additive leaching in barnacles, anemones, Japanese medaka, and an avian physiologically based model (Schmaltz et al. 2020). This poses a potential hazard to human health as individuals may ingest an estimated 39,000–52,000 microplastic particles annually through food and beverages (Cox et al. 2019).

8.1.2 Marine Plastic Management Strategies

According to MacArthur et al. (2017), without effective waste management and infrastructural upgrades, by 2025, the oceans will be flooded with a ton of plastic for every 3 tons of fish. It is expected to contain more plastics than fish by 2050 (MacArthur 2017). Efficient marine waste management encompasses robust collection, separation, and disposal systems, both on land and at sea. Initiatives, such as shoreline cleanups and the establishment of waste collection infrastructure in coastal regions, are integral to stemming the flow of marine debris. Concurrently, strict regulations and active enforcement against illegal dumping are imperative to limit marine pollution. However, marine plastic pollution is hard to address through domestic legislation since plastic cannot be traced across political and geographical boundaries due to ocean currents and prevailing winds. Nevertheless, despite the global distribution issue, there is a need for the development and stricter enforcement of international rules to limit potential hazards to human health and marine ecosystems (McNicholas and Cotton 2019).

Recycling is emerging as a transformative force in sustainable marine waste management (Dąbrowska et al. 2021; Huang et al. 2022). Technological advancements are driving the development of innovative recycling techniques, aiming to repurpose marine plastics into valuable products, ranging from clothing to construction materials (Zhao et al. 2022). Furthermore, circular economy principles, emphasizing the reduction of waste through design and reuse strategies, are gaining prominence in reshaping marine waste management practices (Johansen et al. 2022). Besides, a rise in plastic recycling has the potential to generate more employment opportunities, diminish greenhouse gas emissions, and contribute to advancements in energy efficiency (Hopewell, Dvorak and Kosior 2009; Liu et al. 2018).

The integration of cutting-edge technologies is instrumental in propelling marine waste recycling initiatives (Kumar et al. 2023). Processes like pyrolysis and depolymerization show promise in breaking down complex marine plastics into valuable resources, contributing to the evolution of a circular economy (Roy et al. 2021). The recovered marine plastic wastes can be transformed into value-added petrochemicals, such as aromatic char, hydrogen, synthesis gas, and bio-crude oil, through diverse novel technologies, including thermochemical processes, catalytic conversion, and chemolysis (Kumar et al. 2023; Ng et al. 2023). These technologies not only offer solutions to the challenges posed by diverse waste compositions but also foster a shift toward a more sustainable and resource-efficient approach.

International collaboration stands as a gold standard in tackling the global dimensions of marine waste challenges. As the abundance of transboundary marine litter increases all around the world, this in turn has triggered the development of marine litter policy action plans and interregional collaboration (Graham 2022; Stöfen-O'Brien et al. 2022). Thus, cooperative efforts, agreements, and shared initiatives between nations are crucial for addressing the transboundary nature of marine pollution (Frantzi et al. 2021). Collaborative research and information exchange will enhance the collective understanding of marine waste sources, pathways, and impacts, laying the groundwork for more effective global waste management strategies (Figure 8.2) (Tilbrook et al. 2019; Bettencourt, Costa and Caeiro 2021; Cesarano et al. 2021).

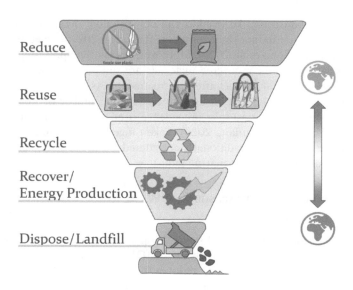

FIGURE 8.2 Marine litter management approaches.

Public engagement remains pivotal for the success of marine waste management initiatives. Community-based programs, educational campaigns, and outreach efforts play a vital role in instilling a sense of responsibility and environmental stewardship (Bettencourt, Costa and Caeiro 2021; Williams and Rangel-Buitrago 2019). Active involvement of local communities in cleanup activities and recycling programs fosters a shared commitment to preserving marine ecosystems.

8.2 UPCYCLING OF MARINE RESOURCES AND BY-PRODUCTS

8.2.1 Principles of Upcycling

The phrase 'upcycling' first appeared in the 1990s and in general upcycling can be defined as 'reuse (discarded objects or material) in such a way as to create a product of higher quality or value than the original' (Bridgens et al. 2018; Wegener 2016). Recently, upcycling has come to be known as an umbrella term encompassing 'creative' kinds of repair, reuse, repurpose, refurbishment, upgrade, remanufacture, and recycling (Sung, Cooper and Kettley 2018). Upcycling is one of the understudied yet potentially eco-friendly behaviors that has the potential to help reduce waste and greenhouse gas emissions (Sung, Cooper and Kettley 2019b). In general, upcycling activities are impacted by the social, economic, and political context in which they take place (Sung, Cooper and Kettley 2019b).

8.2.2 Applications of Upcycling across Industries

The versatility of upcycled marine materials is evident in their diverse applications across industries. From simple household items to more advanced medical equipment or in industries like fashion and construction to art and design, these materials find new life in innovative products (Editorial 2019; Lee et al. 2023; Lee, Lee and Park 2022; Sohn et al. 2020). Moreover, during the upcycling process, the quality is enhanced, so the recovered materials are suitable to be used in the same application as the original product after the treatments. As a result, in order to retain the quality of the material and ensure the maximum number of material cycles, upcycling of plastic trash must take precedence over alternative treatment choices (Horodytska, Kiritsis and Fullana 2020). Taken together, upcycling not only adds economic value to materials traditionally considered waste but also fosters creativity and entrepreneurship within the marine industry.

8.2.3 ENVIRONMENTAL STEWARDSHIP AND WASTE REDUCTION WITH UPCYCLING PRINCIPLES

Upcycling represents environmental stewardship by providing an alternative to conventional waste disposal practices (Roy et al. 2021). By transforming marine plastic waste into valuable goods, upcycling reduces the environmental impact associated with marine-related industries (Mihai et al. 2022). This shift toward waste reduction underscores the importance of responsible resource management.

8.2.4 COMMUNITY ENGAGEMENT AND EDUCATION

The success of upcycling initiatives often hinges on collaboration between communities, industries, and innovators. Engaging stakeholders in the upcycling process fosters a sense of shared responsibility for environmental sustainability (Sung 2021; Sung, Cooper and Kettley 2019a). Educational initiatives focused on upcycling contribute to raising awareness about the environmental consequences of marine activities, promoting a culture of responsible resource use.

8.2.5 CHALLENGES ASSOCIATED WITH UPCYCLING AND FUTURE OUTLOOK

While upcycling presents a promising avenue for sustainable resource management, challenges such as scalability, technological advancements, and market acceptance remain (Balu, Dutta, and Roy Choudhury 2022). To overcome these challenges requires concerted efforts from researchers, industries, and policymakers. The future outlook of upcycling in the marine sector holds great potential for continued innovation and positive environmental impact.

8.3 SUSTAINABILITY AND MARINE BIOTECHNOLOGY

The intersection of sustainability and marine biotechnology represents a frontier of innovation, offering promising solutions to global challenges. As the world struggles with environmental concerns, the sustainable utilization of marine resources through biotechnological advancements emerges as a key avenue. The field of marine biotechnology is centered mostly on the European Union, North America, and Asia-Pacific. China, Japan, Korea, the United States, and Australia are home to some of the world's most known marine biotechnology centers and institutes. In addition, nations such as Middle East, South America, and Africa also actively work in the field of marine biotechnology (Rotter et al. 2021). However, speeding up the production, availability of quality raw materials, lack of investments/funds, lack of appropriate infrastructure, and human capital are becoming major challenges for further expansion of marine biotechnology (Lauritano and Ianora 2018; Rotter et al. 2021).

8.3.1 MAJOR SECTORS CAN BENEFIT FROM MARINE BIOTECHNOLOGY

Marine biotechnology involves the application of biological techniques to harness marine organisms for diverse purposes. From pharmaceuticals and bioenergy to aquaculture and environmental monitoring, marine biotechnology holds the potential of sustainable solutions (Rasmussen and Morrissey 2007). The field of marine biotechnology not only addresses urgent societal needs but also contributes to the conservation of marine ecosystems (Blasiak et al. 2023).

8.3.1.1 Applications of Marine Biotechnology in Sustainable Aquaculture

Global aquaculture production has increased continuously during the last few decades (Naylor et al. 2021). In the 21st century, heavily populated nations such as China and India face significant challenges concerning population, resources, and the environment. Efforts have been made to advance high-quality, efficient, and sustainable blue agriculture using modern biotechnology (Naylor et al. 2021; Xiang 2015). Innovations in breeding techniques, disease management, and nutrition contribute to the development of eco-friendly aquaculture practices (Hoseinifar et al. 2020; Midhun and Arun 2023). These practices not only enhance food security but also minimize the environmental impact associated with traditional aquaculture methods (Kalogerakis et al. 2015).

8.3.1.2 Pharmaceutical Discoveries from Marine Organisms

The ocean, often referred to as the planet's medicine cabinet, harbors a plethora of bioactive compounds with potential pharmaceutical applications (Gates 2010). Marine biotechnology facilitates the exploration and extraction of these compounds, offering new avenues for drug discovery. Compared to the bioactive compounds from terrestrial organisms, marine bioactive compounds are found to possess a higher incidence of significant bioactivity and chemical novelty, which makes marine macroorganisms and microorganisms important sources of novel drug discovery activities (Zhang et al. 2016). Natural drug discovery from marine organisms hence are of greater efficacy and specificity for the treatment of a number of diseases such as cancer, diabetes, obesity, and chronic inflammatory diseases (Sanjeewa et al. 2018). During the past decades, marine natural drug discovery as a science has matured significantly and several drugs developed from marine organisms have been approved by the US FDA or European Medicines Agency (Martins et al. 2014).

In addition, to ensure continuous supply of raw materials, the sustainable harvesting and cultivation of marine bioresources for industrial applications underscores the importance of responsible biotechnological practices.

8.3.1.3 Environmental Monitoring and Remediation

Marine biotechnology plays a crucial role in monitoring and remediating environmental pollution (Aliko et al. 2022). Bioremediation strategies utilizing marine organisms such as bacteria, fungi, nematodes, and microalgae help mitigate the impact of heavy metals, hydrocarbons, plastic, etc. (Das et al. 2009; Fulke, Kotian and Giripunje 2020; Mohanrasu et al. 2020). Monitoring tools, powered by biotechnological innovations, provide real-time data to assess and address ecological threats, contributing to sustainable marine conservation efforts. Despite the promising potential, marine biotechnology requires addressing several challenges to ensure the sustainability of the blue bioeconomy. Striking a balance between exploitation and conservation, addressing ethical concerns related to genetic modification, and ensuring equitable access to biotechnological benefits are critical aspects that demand attention as this field progresses.

8.4   ESTABLISHING CIRCULAR SUPPLY CHAINS FOR SUSTAINABLE MARINE RESOURCE MANAGEMENT

As the world struggles with complex challenges faced by blue bioeconomy rated industries such as climate change, overfishing, and environmental degradation, the imperative for innovative solutions in the management of marine resources becomes increasingly urgent. This section briefly discusses the transformative potential of circular supply chains within the marine industry and delves into the multifaceted strategies that can revolutionize the sustainable management of marine resources.

8.4.1   WHAT IS CIRCULAR SUPPLY CHAIN?

The circular economy is gaining growing recognition as a superior alternative to the prevalent linear economic model, characterized by a take, make, and dispose approach (Farooque et al. 2019). At its core, circular supply chains represent an ecological paradigm shift, emphasizing the necessity of closing resource loops to minimize waste and environmental impact and promote natural regeneration under the circular economy concept (Theeraworawit, Suriyankietkaew and Hallinger 2022). In the context of marine resource management, this holistic approach not only prioritizes sustainable practices but also integrates cutting-edge technologies and community engagement, creating a comprehensive framework for ecological stewardship.

8.4.2   KEY COMPONENTS OF CIRCULAR SUPPLY CHAINS IN MARINE RESOURCE MANAGEMENT

8.4.2.1 Sustainable Fishing Practices

Preserving fish stocks and minimizing bycatch is key to the sustainable fisheries. However, it requires a complex and more practical approach that combines technological innovation with ecological awareness (Little et al. 2014). Technologies such as GPS tracking and real-time monitoring not only enhance the efficiency of fishing operations but also contribute to adaptive management strategies (Glaviano et al. 2022).

Besides, the integration of ecosystem-based approaches such as ecosystem-based management (EBM) for the fishing industry ensures that fishing practices align with the broader goal of maintaining marine biodiversity (Hegland, Raakjær and van Tatenhove 2015). However, the lack of success in managing marine fish resources is linked to the narrow focus of traditional management on distinct sectors. Activities like shipping, energy exploration, fisheries, tourism, and coastal development have traditionally been treated separately, ignoring the combined effects on marine ecosystems, such as pollution, eutrophication, and habitat destruction. Addressing these impacts in isolation is considered inadequate, resulting in a fragmented perspective on resource management. With an increasing recognition of the interconnected nature of ecosystems, EBM approaches look promising to ensure the sustainability of the fishing industry (Curtin and Prellezo 2010).

8.4.2.2　Resource Recovery and Recycling

Efficient processing and recycling of by-products from marine industries significantly contribute to waste reduction. According to previous studies, a considerable proportion of the total fish captured is not consumed and is discarded without further processing. Generally, fish tissue that is not suitable for consumption, bones, guts, gills, and fins are on the discarding list (Jayathilakan et al. 2012; Marti-Quijal et al. 2020). According to earlier studies, these fish waste still contains 15–30% crude protein, 0–25% fat, minerals, collagen, and its derivatives, and chitin (Nawaz et al. 2020). Thus, most of the resulting waste can be used to produce different value-added products such as fishmeal, liquid fertilizer, chitosan, proteins, and enzymes (Chandra and Shamasundar 2011; Jayathilakan et al. 2012; Kristbergsson and Arason 2007). Moreover, innovative technologies, such as enzymatic digestion and microbial fermentation, can transform fish waste into high-value products, ranging from nutraceuticals to biofuels (Marti-Quijal et al. 2020). Circular supply chains also necessitate a shift toward sustainable packaging materials, fostering a closed-loop system that minimizes the environmental impact of marine-related industries. However, the utilization of fish by-products in human foods is still challenging with respect to safety issues and their interactions with other ingredients in foods (Nawaz et al. 2020).

8.4.2.3　Traceability and Transparency

Advanced tracking systems, including blockchain technology, not only ensure transparency in the supply chain but also empower consumers to make informed choices (Dutta et al. 2020). By providing a detailed account of the journey from ocean to plate, traceability measures instill confidence in consumers who increasingly seek ethically and sustainably sourced marine products (Belton, Reardon and Zilberman 2020; Freitas, Vaz-Pires and Câmara 2020; Razak, Hendry and Stevenson 2021). Transparent supply chains also act as a barrier to illegal, unreported, and unregulated fishing practices (Ismail et al. 2023).

8.4.2.4　Community Engagement

Circular supply chains must prioritize collaboration with local communities, recognizing the importance of traditional knowledge (TK) and practices.

Specifically, TK develops within indigenous societies through generations of learning, gaining experience in the surroundings, and adapting to the native environment. TK is passed down from one generation to another and TK is a critical factor in achieving sustainable development and growth of the global economy (Kumari et al. 2022). Furthermore, beyond economic considerations, community engagement involves education initiatives, capacity-building, and the incorporation of indigenous wisdom into sustainable resource management practices (Valente 2012). Empowering local stakeholders enhances the resilience of marine ecosystems and ensures the long-term sustainability of circular supply chain initiatives.

8.4.3 BENEFITS OF CIRCULAR SUPPLY CHAINS IN MARINE RESOURCE MANAGEMENT

8.4.3.1 Resource Conservation

By minimizing waste and promoting efficient resource use, circular supply chains contribute significantly to the conservation of marine ecosystems and biodiversity (Ward et al. 2022). Sustainable practices not only preserve the delicate balance of marine ecosystems but also foster the rehabilitation of degraded areas, ensuring the longevity of diverse marine species and habitats (Ward et al. 2022).

8.4.3.2 Economic Viability

Circularity in the marine industry generates economic opportunities through innovation and entrepreneurship. Beyond the development of recycling technologies, sustainable aquaculture ventures, and circular economy practices, there is potential for the creation of green jobs, supporting local economies, and fostering resilience in the face of economic uncertainties (Dantas et al. 2021; Suchek, Ferreira and Fernandes 2022).

8.4.3.3 Climate Resilience

Circular supply chains play a crucial role in climate resilience by reducing the carbon footprint associated with traditional linear models (Tiwari et al. 2023). The conservation of marine ecosystems through sustainable practices aids in mitigating the impacts of climate change on vulnerable coastal communities, ensuring the continued provision of ecosystem services such as shoreline protection and carbon sequestration (Naumann et al. 2011).

8.4.3.4 Enhanced Reputation

Companies embracing circular supply chains gain a competitive advantage by demonstrating a robust commitment to sustainability. Transparent and ethical practices, coupled with a dedication to reducing environmental impact, contribute to building a positive brand image. This not only attracts environmentally conscious consumers but also fosters partnerships with like-minded entities, creating a network of organizations committed to ecological stewardship.

8.5   FUTURE DIRECTIONS

While the concept of circular supply chains holds immense promise, several challenges persist, including technological barriers, regulatory frameworks, and initial implementation costs. Collaboration between governments, industries, and research institutions is paramount for overcoming these challenges and fostering widespread adoption. Continued research and development are essential to refine circular supply chain practices, ensuring their adaptability to evolving environmental and economic conditions.

REFERENCES

Editorial. 2019. "Plastic upcycling." *Nature Catalysis* 2 (11):945–946. doi: 10.1038/s41929-019-0391-7

Agamuthu, P., S. B. Mehran, A. Norkhairah, and A. Norkhairiyah. 2019. "Marine debris: A review of impacts and global initiatives." *Waste Management & Research* 37 (10):987–1002. doi: 10.1177/0734242X19845041

Aliko, V., C. R. Multisanti, B. Turani, and C. Faggio. 2022. "Get rid of marine pollution: Bioremediation an innovative, attractive, and successful cleaning strategy." *Sustainability* 14 (18). doi: 10.3390/su141811784

Balu, R., N. K. Dutta, and N. Roy Choudhury. 2022. "Plastic waste upcycling: A sustainable solution for waste management, product development, and circular economy." *Polymers* 14 (22). doi: 10.3390/polym14224788

Belton, B., T. Reardon, and D. Zilberman. 2020. "Sustainable commoditization of seafood." *Nature Sustainability* 3 (9):677–684. doi: 10.1038/s41893-020-0540-7

Bettencourt, S., S. Costa, and S. Caeiro. 2021. "Marine litter: A review of educative interventions." *Marine Pollution Bulletin* 168:112446. doi: 10.1016/j.marpolbul.2021.112446

Blasiak, R., J. B. Jouffray, D. J. Amon, J. Claudet, P. Dunshirn, P. Sogaard Jorgensen, A. Pranindita, C. C. C. Wabnitz, E. Zhivkoplias, and H. Osterblom. 2023. "Making marine biotechnology work for people and nature." *Nature Ecology and Evolution* 7 (4):482–485. doi: 10.1038/s41559-022-01976-9

Bridgens, B., M. Powell, G. Farmer, C. Walsh, E. Reed, M. Royapoor, P. Gosling, J. Hall, and O. Heidrich. 2018. "Creative upcycling: Reconnecting people, materials and place through making." *Journal of Cleaner Production* 189:145–154. doi: 10.1016/j.jclepro.2018.03.317.

Cesarano, C., G. Aulicino, C. Cerrano, M. Ponti, and S. Puce. 2021. "Scientific knowledge on marine beach litter: A bibliometric analysis." *Marine Pollution Bulletin* 173 (Pt B):113102. doi: 10.1016/j.marpolbul.2021.113102

Chandra, M. V., and B. A. Shamasundar. 2011. "Fish Processing Waste Management." In *Food Processing Waste Management: Treatment and Utilization Technology*, 161–194. New Delhi: New India Publishing Agency.

Cox, K. D., G. A. Covernton, H. L. Davies, J. F. Dower, F. Juanes, and S. E. Dudas. 2019. "Human consumption of microplastics." *Environmental Science & Technology* 53 (12):7068–7074. doi: 10.1021/acs.est.9b01517

Cressey, D. 2016. "Bottles, bags, ropes and toothbrushes: The struggle to track ocean plastics." *Nature* 536 (7616):263–265. doi: 10.1038/536263a

Curtin, R., and R. Prellezo. 2010. "Understanding marine ecosystem based management: A literature review." *Marine Policy* 34 (5):821–830. doi: 10.1016/j.marpol.2010.01.003

Dąbrowska, J., M. Sobota, M. Świąder, P. Borowski, A. Moryl, R. Stodolak, E. Kucharczak, Z. Zięba, and J. K. Kazak. 2021. "Marine waste: Sources, fate, risks, challenges and research needs." *International Journal of Environmental Research and Public Health* 18 (2). doi: 10.3390/ijerph18020433

Dantas, T. E. T., E. D. de-Souza, I. R. Destro, G. Hammes, C. M. T. Rodriguez, and S. R. Soares. 2021. "How the combination of circular economy and Industry 4.0 can contribute towards achieving the sustainable development goals." *Sustainable Production and Consumption* 26:213–227. doi: 10.1016/j.spc.2020.10.005

Das, S., A. Elavarasi, P. S. Lyla, and S. A. Khan. 2009. "Biosorption of heavy metals by marine bacteria: Potential tool for detecting marine pollution." *Environmental Health* 9 (1/2):38–43.

Dutta, P., T. M. Choi, S. Somani, and R. Butala. 2020. "Blockchain technology in supply chain operations: Applications, challenges and research opportunities." *Transportation Research Part E: Logistics and Transportation Review* 142:102067. doi: 10.1016/j.tre.2020.102067

Farooque, M., A. Zhang, M. Thürer, T. Qu, and D. Huisingh. 2019. "Circular supply chain management: A definition and structured literature review." *Journal of Cleaner Production* 228:882–900. doi: 10.1016/j.jclepro.2019.04.303

Frantzi, S., R. Brouwer, E. Watkins, P. van Beukering, M. C. Cunha, H. Dijkstra, S. Duijndam, H. Jaziri, I. C. Okoli, M. Pantzar, I. Rada Cotera, K. Rehdanz, K. Seidel, and G. Triantaphyllidis. 2021. "Adoption and diffusion of marine litter clean-up technologies across European seas: Legal, institutional and financial drivers and barriers." *Marine Pollution Bulletin* 170:112611. doi: 10.1016/j.marpolbul.2021.112611.

Freitas, J., P. Vaz-Pires, and J. S. Câmara. 2020. "From aquaculture production to consumption: Freshness, safety, traceability and authentication, the four pillars of quality." *Aquaculture* 518:734857. doi: 10.1016/j.aquaculture.2019.734857

Fulke, A. B., A. Kotian, and M. D. Giripunje. 2020. "Marine microbial response to heavy metals: Mechanism, implications and future prospect." *Bulletin of Environmental Contamination and Toxicology* 105 (2):182–197. doi: 10.1007/s00128-020-02923-9

Gates, K. W. 2010. "Marine products for healthcare: Functional and bioactive nutraceutical compounds from the ocean, *Vazhiyil venugopal.*" *Journal of Aquatic Food Product Technology* 19 (1):48–54. doi: 10.1080/10498850903517528

Glaviano, F., R. Esposito, A. D. Cosmo, F. Esposito, L. Gerevini, A. Ria, M. Molinara, P. Bruschi, M. Costantini, and V. Zupo. 2022. "Management and sustainable exploitation of marine environments through smart monitoring and automation." *Journal of Marine Science and Engineering* 10 (2). doi: 10.3390/jmse10020297

Graham, R. E. D. 2022. "Achieving greater policy coherence and harmonisation for marine litter management in the North-East Atlantic and wider Caribbean region." *Marine Pollution Bulletin* 180:113818. doi: 10.1016/j.marpolbul.2022.113818

Hegland, T. J., J. Raakjær, and J. van Tatenhove. 2015. "Implementing ecosystem-based marine management as a process of regionalisation: Some lessons from the Baltic Sea." *Ocean & Coastal Management* 117:14–22. doi: 10.1016/j.ocecoaman.2015.08.005

Hofmann, T., S. Ghoshal, N. Tufenkji, J. F. Adamowski, S. Bayen, Q. Chen, P. Demokritou, M. Flury, T. Hüffer, N. P. Ivleva, R. Ji, R. L. Leask, M. Maric, D. M. Mitrano, M. Sander, S. Pahl, M. C. Rillig, T. R. Walker, J. C. White, and K. J. Wilkinson. 2023. "Plastics can be used more sustainably in agriculture." *Communications Earth & Environment* 4 (1):332. doi: 10.1038/s43247-023-00982-4

Hopewell, J., R. Dvorak, and E. Kosior. 2009. "Plastics recycling: Challenges and opportunities." *Philosophical Transactions of the Royal Society of London* 364 (1526):2115–26. doi: 10.1098/rstb.2008.0311

Horodytska, O., D. Kiritsis, and A. Fullana. 2020. "Upcycling of printed plastic films: LCA analysis and effects on the circular economy." *Journal of Cleaner Production* 268:122138. doi: 10.1016/j.jclepro.2020.122138

Hoseinifar, S. H., Y.-Z. Sun, Z. Zhou, H. Van Doan, S. J. Davies, and R. Harikrishnan. 2020. "Boosting immune function and disease bio-control through environment-friendly and sustainable approaches in finfish aquaculture: Herbal therapy scenarios." *Reviews in Fisheries Science & Aquaculture* 28 (3):303–321. doi: 10.1080/23308249.2020.1731420

Huang, J., A. Veksha, W. P. Chan, A. Giannis, and G. Lisak. 2022. "Chemical recycling of plastic waste for sustainable material management: A prospective review on catalysts and processes." *Renewable and Sustainable Energy Reviews* 154:111866. doi: 10.1016/j.rser.2021.111866

Ismail, S., H. Reza, K. Salameh, H. Kashani Zadeh, and F. Vasefi. 2023. "Toward an intelligent blockchain IoT-enabled fish supply chain: A review and conceptual framework." *Sensors* 23 (11). doi: 10.3390/s23115136

IUCN (International Union for Conservation of Nature). 2021. "IUCN publications on marine plastic pollution." https://www.iucn.org/sites/default/files/2022-04/marine_plastic_pollution_issues_brief_nov21.pdf (accessed December 31, 2023).

Jayathilakan, K., K. Sultana, K. Radhakrishna, and A. S. Bawa. 2012. "Utilization of byproducts and waste materials from meat, poultry and fish processing industries: A review." *Journal of Food Science and Technology* 49 (3):278–93. doi: 10.1007/s13197-011-0290-7

Johansen, M. R., T. B. Christensen, T. M. Ramos, and K. Syberg. 2022. "A review of the plastic value chain from a circular economy perspective." *Journal of Environmental Management* 302 (Pt A):113975. doi: 10.1016/j.jenvman.2021.113975

Kalogerakis, N., J. Arff, I. M. Banat, O. J. Broch, D. Daffonchio, T. Edvardsen, H. Eguiraun, L. Giuliano, A. Handa, K. Lopez-de-Ipina, I. Marigomez, I. Martinez, G. Oie, F. Rojo, J. Skjermo, G. Zanaroli, and F. Fava. 2015. "The role of environmental biotechnology in exploring, exploiting, monitoring, preserving, protecting and decontaminating the marine environment." *New Biotechnology* 32 (1):157–67. doi: 10.1016/j.nbt.2014.03.007

Kristbergsson, K., and S. Arason. 2007. "Utilization of By-Products in the Fish Industry." In *Utilization of By-Products and Treatment of Waste in the Food Industry*, edited by V. Oreopoulou and W. Russ, vol. 3. Boston, MA: Springer.

Kumar, M., S. Bolan, L. P. Padhye, M. Konarova, S. Y. Foong, S. S. Lam, S. Wagland, R. Cao, Y. Li, N. Batalha, M. Ahmed, A. Pandey, K. H. M. Siddique, H. Wang, J. Rinklebe, and N. Bolan. 2023. "Retrieving back plastic wastes for conversion to value added petrochemicals: Opportunities, challenges and outlooks." *Applied Energy* 345:121307. doi: 10.1016/j.apenergy.2023.121307

Kumari, M., S. Pandey, V. P. Giri, P. Chauhan, N. Mishra, P. Verma, A. Tripathi, S. P. Singh, R. Bajpai, and A. Mishra. 2022. "Integrated Approach for Technology Transfer Awareness of Traditional Knowledge for Upliftment of Circular Bioeconomy." In *Biomass, Biofuels, Biochemicals*, edited by S. Varjani, A. Pandey, T. Bhaskar, S. V. Mohan, and D. C. W. Tsang, 613–636. Elsevier.

Laskar, N., and U. Kumar. 2019. "Plastics and microplastics: A threat to environment." *Environmental Technology & Innovation* 14:100352. doi: 10.1016/j.eti.2019.100352

Lauritano, C., and A. Ianora. 2018. "Grand Challenges in Marine Biotechnology: Overview of Recent EU-Funded Projects." In *Grand Challenges in Marine Biotechnology*, edited by P. H. Rampelotto and A. Trincone, 425–449. Cham: Springer International Publishing.

Lebreton, L. C. M., J. van der Zwet, J. W. Damsteeg, B. Slat, A. Andrady, and J. Reisser. 2017. "River plastic emissions to the world's oceans." *Nature Communications* 8:15611. doi: 10.1038/ncomms15611

Lee, S., Y. R. Lee, S. J. Kim, J.-S. Lee, and K. Min. 2023. "Recent advances and challenges in the biotechnological upcycling of plastic wastes for constructing a circular bioeconomy." *Chemical Engineering Journal* 454:140470. doi: 10.1016/j.cej.2022.140470

Lee, S., J. Lee, and Y.-K. Park. 2022. "Simultaneous upcycling of biodegradable plastic and sea shell wastes through thermocatalytic monomer recovery." *ACS Sustainable Chemistry & Engineering* 10 (42):13972–13979. doi: 10.1021/acssuschemeng.2c04050

Lestari, P., and Y. Trihadiningrum. 2019. "The impact of improper solid waste management to plastic pollution in Indonesian coast and marine environment." *Marine Pollution Bulletin* 149:110505. doi: 10.1016/j.marpolbul.2019.110505

Little, A. S., C. L. Needle, R. Hilborn, D. S. Holland, and C. T. Marshall. 2014. "Real-time spatial management approaches to reduce bycatch and discards: Experiences from Europe and the United States." *Fish and Fisheries* 16 (4):576–602. doi: 10.1111/faf.12080

Liu, Z., M. Adams, R. P. Cote, Q. Chen, R. Wu, Z. Wen, W. Liu, and L. Dong. 2018. "How does circular economy respond to greenhouse gas emissions reduction: An analysis of Chinese plastic recycling industries." *Renewable and Sustainable Energy Reviews* 91:1162–1169. doi: 10.1016/j.rser.2018.04.038

MacArthur D. E. 2017. "Beyond plastic waste." *Science* 358:843–843.doi: 10.1126/science. aao6749

Martins, A., H. Vieira, H. Gaspar, and S. Santos. 2014. "Marketed marine natural products in the pharmaceutical and cosmeceutical industries: Tips for success." *Marine Drugs* 12 (2):1066–1101. doi: 10.3390/md12021066

Marti-Quijal, F. J., F. Remize, G. Meca, E. Ferrer, M.-J. Ruiz, and F. J. Barba. 2020. "Fermentation in fish and by-products processing: An overview of current research and future prospects." *Current Opinion in Food Science* 31:9–16. doi: 10.1016/j. cofs.2019.08.001

McNicholas, G., and M. Cotton. 2019. "Stakeholder perceptions of marine plastic waste management in the United Kingdom." *Ecological Economics* 163:77–87. doi: 10.1016/j. ecolecon.2019.04.022

Midhun, S. J., and D. Arun. 2023. "History and Development of Microbial Technology in Aquaculture." In *Recent Advances in Aquaculture Microbial Technology*, edited by J. Mathew, M. S. Jose, Radhakrishnan E. K and Ajay Kumar, 1–13. Academic Press.

Mihai, F.-C., S. Gündoğdu, F. R. Khan, A. Olivelli, L. A. Markley, and T. van Emmerik. 2022. "Plastic Pollution in Marine and Freshwater Environments: Abundance, Sources, and Mitigation." In *Emerging Contaminants in the Environment*, edited by H. Sarma, D. C. Dominguez, and W.-Y. Lee, 241–274. Elsevier.

Mohanrasu, K., R. G. R. Rao, R. Raja, and A. Arun. 2020. "Bioremediation Process by Marine Microorganisms." In Encyclopedia of Marine Biotechnology, edited by S.-K. Kim, 2211–2228. Wiley-Blackwell.

Naumann, S., G. Anzaldua, P. Berry, S. Burch, M. Davis, A. Frelih-Larsen, H. Gerdes, and M. Sanders. 2011. "Assessment of the potential of ecosystem-based approaches to climate change adaptation and mitigation in Europe." *Final Report to the European Commission, DG Environment.*

Nawaz, A., E. Li, S. Irshad, Z. Xiong, H. Xiong, H. M. Shahbaz, and F. Siddique. 2020. "Valorization of fisheries by-products: Challenges and technical concerns to food industry." *Trends in Food Science & Technology* 99:34–43. doi: 10.1016/j.tifs.2020.02.022

Naylor, R. L., R. W. Hardy, A. H. Buschmann, S. R. Bush, L. Cao, D. H. Klinger, D. C. Little, J. Lubchenco, S. E. Shumway, and M. Troell. 2021. "A 20-year retrospective review of global aquaculture." *Nature* 591 (7851):551–563. doi: 10.1038/ s41586-021-03308-6

Ng, K. W. J., J. S. K. Lim, N. Gupta, B. X. Dong, C. P. Hu, J. Hu, and X. M Hu. 2023. "A facile alternative strategy of upcycling mixed plastic waste into vitrimers." *Communications Chemistry* 6 (1):158. doi: 10.1038/s42004-023-00949-8.

Okoffo, E. D., E. Donner, S. P. McGrath, B. J. Tscharke, J. W. O'Brien, S. O'Brien, F. Ribeiro, S. D. Burrows, T. Toapanta, C. Rauert, S. Samanipour, J. F. Mueller, and K. V. Thomas. 2021. "Plastics in biosolids from 1950 to 2016: A function of global plastic production and consumption." *Water Research* 201:117367. doi: 10.1016/j. watres.2021.117367

Phelan, A. A., H. Ross, N. A. Setianto, K. Fielding, and L. Pradipta. 2020. "Ocean plastic crisis-mental models of plastic pollution from remote Indonesian coastal communities." *PLoS One* 15 (7):e0236149. doi: 10.1371/journal.pone.0236149

Pradeep, S. A., R. K. Iyer, H. Kazan, and S. Pilla. 2017. "Automotive Applications of Plastics: Past, Present, and Future." In *Applied Plastics Engineering Handbook*, edited by M. Kutz, 651–673. William Andrew Publishing.

Rasmussen, R. S., and M. T. Morrissey. 2007. "Marine Biotechnology for Production of Food Ingredients." In *Advances in Food and Nutrition Research*, 237–292. Academic Press.

Razak, G. M., L. C. Hendry, and M. Stevenson. 2021. "Supply chain traceability: A review of the benefits and its relationship with supply chain resilience." *Production Planning & Control* 34 (11):1114–1134. doi: 10.1080/09537287.2021.1983661

Rotter, A., et al. 2021. "The essentials of marine biotechnology." *Frontiers in Marine Science* 8:629629.

Roy, P. S., G. Garnier, F. Allais, and K. Saito. 2021. "Strategic approach towards plastic waste valorization: Challenges and promising chemical upcycling possibilities." *ChemSusChem* 14 (19):4007–4027. doi: 10.1002/cssc.202100904

Sangroniz, A., J. B. Zhu, X. Tang, A. Etxeberria, E. Y. Chen, and H. Sardon. 2019. "Packaging materials with desired mechanical and barrier properties and full chemical recyclability." *Nature Communications* 10 (1):3559. doi: 10.1038/s41467-019-11525-x

Sanjeewa, K. K. A., N. Kang, G. Ahn, Y. Jee, Y.-T. Kim, and Y.-J. Jeon. 2018. "Bioactive potentials of sulfated polysaccharides isolated from brown seaweed *Sargassum* spp in related to human health applications: A review." *Food Hydrocolloids* 81:200–208. doi: 10.1016/j.foodhyd.2018.02.040.

Santos, M. R., L. C. Dias, M. C. Cunha, and J. R. Marques. 2022. "Multicriteria decision analysis addressing marine and terrestrial plastic waste management: A review." *Frontiers in Marine Science* 8. doi: 10.3389/fmars.2021.747712

Schmaltz, E., E. C. Melvin, Z. Diana, E. F. Gunady, D. Rittschof, J. A. Somarelli, J. Virdin, and M. M. Dunphy-Daly. 2020. "Plastic pollution solutions: Emerging technologies to prevent and collect marine plastic pollution." *Environment International* 144:106067. doi: 10.1016/j.envint.2020.106067

Sohn, Y. J., H. T. Kim, K. A. Baritugo, S. Y. Jo, H. M. Song, S. Y. Park, S. K. Park, J. Pyo, H. G. Cha, H. Kim, J. G. Na, C. Park, J. I. Choi, J. C. Joo, and S. J Park. 2020. "Recent advances in sustainable plastic upcycling and biopolymers." *Biotechnology Journal* 15 (6):e1900489. doi: 10.1002/biot.201900489

Statista. 2023. "Annual production of plastics worldwide from 1950 to 2021." https://www.statista.com/statistics/282732/global-production-of-plastics-since-1950/

Stöfen-O'Brien, A., A. Naji, A. L. Brooks, J. R. Jambeck, and F. R. Khan. 2022. "Marine plastic debris in the Arabian/Persian Gulf: Challenges, opportunities and recommendations from a transdisciplinary perspective." *Marine Policy* 136:104909. doi: 10.1016/j.marpol.2021.104909

Suchek, N., J. J. Ferreira, and P. O. Fernandes. 2022. "A review of entrepreneurship and circular economy research: State of the art and future directions." *Business Strategy and the Environment* 31 (5):2256–2283. doi: 10.1002/bse.3020

Sung, K. 2021. "Evaluating two interventions for scaling up upcycling: Community event and upcycling plaza." *IOP Conference Series: Materials Science and Engineering* 1196 (1):012001. doi: 10.1088/1757-899x/1196/1/012001

Sung, K., T. Cooper, and S. Kettley. 2018. "Emerging Social Movements for Sustainability: Understanding and Scaling Up Upcycling in the UK." In *The Palgrave Handbook of Sustainability*, edited by R. Brinkmann and S. J. Garren, 299–312. Cham: Springer International Publishing.

Sung, K., T. Cooper, and S. Kettley. 2019a. "Developing interventions for scaling up UK upcycling." *Energies* 12 (14). doi: 10.3390/en12142778.

Sung, K., T. Cooper, and S. Kettley. 2019b. "Factors influencing upcycling for UK makers." *Sustainability* 11 (3). doi: 10.3390/su11030870

Theeraworawit, M., S. Suriyankietkaew, and P. Hallinger. 2022. "Sustainable supply chain management in a circular economy: A bibliometric review." *Sustainability* 14 (15). doi: 10.3390/su14159304.

Tilbrook, et al. 2019. "An enhanced ocean acidification observing network: From people to technology to data synthesis and information exchange." *Frontiers in Marine Science* 6. doi: 10.3389/fmars.2019.00337

Tiwari, S., K. S. Mohammed, G. Mentel, S. Majewski, and I. Shahzadi. 2023. "Role of circular economy, energy transition, environmental policy stringency, and supply chain pressure on CO_2 emissions in emerging economies." *Geoscience Frontiers*:101682. doi: 10.1016/j.gsf.2023.101682

Valente, M. 2012. "Indigenous resource and institutional capital." *Business & Society* 51 (3):409–449. doi: 10.1177/0007650312446680

Ward, D., J. Melbourne-Thomas, G. T. Pecl, K. Evans, M. Green, P. C. McCormack, C. Novaglio, R. Trebilco, N. Bax, M. J. Brasier, E. L. Cavan, G. Edgar, H. L. Hunt, J. Jansen, R. Jones, M. A. Lea, R. Makomere, C. Mull, J. M. Semmens, J. Shaw, D. Tinch, T. J. van Stevenick, and C. Layton. 2022. "Safeguarding marine life: Conservation of biodiversity and ecosystems." *Reviews in Fish Biology and Fisheries* 32 (1):65–100. doi: 10.1007/s11160-022-09700-3

Wegener, C. 2016. "Upcycling." In *Creativity: A New Vocabulary*, edited by V. P. Glăveanu, L. Tanggaard, and C. Wegener, 181–188. London: Palgrave Macmillan.

Williams, A. T., and N. Rangel-Buitrago. 2019. "Marine litter: Solutions for a major environmental problem." *Journal of Coastal Research* 35 (3):648–663. doi: 10.2112/jcoastres-d-18-00096.1

Wu, Y., P. Guo, X. Zhang, Y. Zhang, S. Xie, and J. Deng. 2019. "Effect of microplastics exposure on the photosynthesis system of freshwater algae." *Journal of Hazardous Materials* 374:219–227. doi: 10.1016/j.jhazmat.2019.04.039

Xiang, J. 2015. "Recent major advances of biotechnology and sustainable aquaculture in China." *Current Biotechnology* 4 (3):296–310. doi: 10.2174/2211550105666151105190012

Yuan, Z., R. Nag, and E. Cummins. 2022. "Human health concerns regarding microplastics in the aquatic environment: From marine to food systems." *Science of the Total Environment* 823:153730. doi: 10.1016/j.scitotenv.2022.153730

Zhang, G., J. Li, T. Zhu, Q. Gu, and D Li. 2016. "Advanced tools in marine natural drug discovery." *Current Opinion in Biotechnology* 42:13–23. doi: 10.1016/j.copbio.2016.02.021

Zhao, X., B. Boruah, K. F. Chin, M. Dokic, J. M. Modak, and H. S. Soo. 2022. "Upcycling to sustainably reuse plastics." *Advanced Materials* 34 (25):e2100843. doi: 10.1002/adma.202100843

Index

Note: Locators in *italics* represent figures and **bold** indicate tables in the text.

A

Acanthophora spicifera, 108
Acetone, 33
Acremonium sclerotigenum, 67
Agarases, 41
Agricultural applications of microalgae, 81–82
Alginate, 26–27
Alginic acid, 26–27
α-amylase, 41
Alternaria sp.
    A. alternata, 64
    A. destruens, 63
    A. radicina, 108
    A. solani, 108
3,6-Anhydrogalactose, 27
Anisakis spp., 95
Antibiotics, 41–42
    from marine microorganisms, 111
Anticancer compounds from marine organisms,
    111–112
Anti-inflammatory agents from marine
    organisms, 112
Antioxidants, 107
    from marine organisms, 110
Antiviral properties of marine organisms, 111
Aphanizomenon flos-aquae, 81
Aquaculture
    effects of nematodes in, 95
    marine nematodes in, 93–94
    nutrient cycling within aquaculture
        ponds, 94
    role of marine bacteria in, 41–42
Aquaponics and hydroponics applications,
    107–108
Argyrops spinifer, 98
Arthrospira maxima, 52
Ascomycota, 61
Ascophyllum nodosum, 106, 108
Asexual reproduction in seaweeds, 21
    propagules, 21
    spores, 21–22
    thalli fragmentation, 21
Aspergillus sp.
    A. candidus, 65
    A. terreus, 67
    A. versicolor, 63
Auto-flocculation, 79
Auxiliary propulsion, 6

B

Bacidomycota, 61
Bacillus spp., 49
Bacteria, marine, 40
    application of
        as biosurfactant, 47
        in petroleum and diesel biodegradation,
            46–47
        as plastic-degrading agents, **50**
    industrially useful enzymes, 40–41
    role of
        in aquaculture, 41–42
        in degradation of plastic, 47–50
        in heavy metal bioremediation, 43–46
Bacteriocins, 42, **45**
Bathylaimus sp.
    B. capacosus, 99
    B. tenuicaudatus, 99
Benzodiazepine alkaloids, 113
β-D-glucuronic acid, 28
Biodegradation processes, 43
Bioflocculation, 79
Biofuel production systems, 84–85
Biological microalgae harvesting methods
    bioflocculation, 79
    microbial lysis, 80
    predation, 79
Biomass energy, 7–8
Biomedical industry, 3–4
Biomolecules, 65
Bionic prospecting, 10
Bioprospecting, marine, 8
    bionic prospecting, 10
    chemical prospecting, 9–10
    gene prospecting, 10
Bioremediation, 43
    of marine nematodes in polluted
        environments, 97–99
Bioresources, marine, 105
    bioactive marine compounds roles in
        functional food industry, 109
    antioxidant properties, 110
    cosmeceuticals, 113
    future directions, 115
    gelatin separated from marine fish
        by-products, 113
    biomedical applications of fish
        gelatin, 115

food industry applications of gelatin,
114–115
general extraction protocol of fish
gelatin, 114
medicinal value of, 110
antibiotics, 111
anticancer compounds, 111–112
anti-inflammatory agents, 112
antiviral properties of marine organisms,
111
cardiovascular medications, 112–113
in new plant fertilizer products, 105, *106*
aquaponics and hydroponics applications,
107–108
biodegradable mulching materials, 108
biostimulation properties reported from
marine plant fertilizers, 106
disease resistance and plant immunity,
108, *109*
soil salinity stress, alleviation of, 107
stimulating root growth and nutrient
uptake, 108
Biosurfactant, applications of marine bacteria
as, **47**
Biotechnological potential of marine
nematodes, 97
Biotechnology, 1–2
Blue bioeconomy, defined, 2
Blue biotechnology, 1, **2**
Blue-green algae, *see* Cyanobacteria, marine
Blue Tourism, 4
Botrytis cinerea, 108
Brazilian Exclusive Economic Zone, 11
Brown biotechnology, 1, **2**
Brown seaweeds, bioactive phlorotannin reported
from, *31*

C

CAGR, *see* Compound annual growth rate
Calcium carbonate, 11
Cardiovascular medications from marine
organisms, 112–113
Carotenoids, 31–32, 113
Carpogonium, 22
Carrageenan, 27
Carrageenases, 41
Caulerpa serrulata, 108
Cellulose-degrading enzymes, 41
Centrifugation technique, 78
Chemical microalgae harvesting methods, 79
auto-flocculation, 79
coagulation, 79
Chemical prospecting, 9–10
Chitin, 65
Chitinases, 41
Chlorella sp., 81, 84

Chlorophyll *a*, 22
Chlorophyll *b*, 22
Cholesterols, 32
Chondrus crispus, 27
Chytridiomycota, 61
Circular economy principles related to blue
bioeconomy, 122
circular supply chain, 128
future directions, 131
marine waste management and recycling, 122
marine plastic management strategies,
124–125
plastic as marine contaminant, 122–123,
123
sustainability and marine biotechnology, 126
environmental monitoring and
remediation, 128
marine biotechnology applications in
sustainable aquaculture, 127
pharmaceutical discoveries from marine
organisms, 127
upcycling
applications of, 125–126
challenges associated with, 126
community engagement and
education, 126
environmental stewardship and waste
reduction with, 126
principles of, 125
Circular supply chains in marine resource
management, 128
benefits of
climate resilience, 130
economic viability, 130
enhanced reputation, 130
resource conservation, 130
key components of
community engagement, 129–130
resource recovery and recycling, 129
sustainable fishing practices, 128–129
traceability and transparency, 129
Cladosporium sp., 63
Climate resilience, 130
Coagulation, 79
Coastal tourism, 4, 5
maritime parks, tours to, 5
scuba diving, 4
spearfishing, 4–5
water skiing, 5
wildlife mammal watching, 5
Codium taylorii, 108
Collagenases, 41
Collectively bioprospecting, defined, 9
Colors of biotechnology, 1–2, **2**
Community engagement, 129–130
Compound annual growth rate (CAGR), 73
Corollospora lacera, 65

Cosmeceuticals from marine organisms, 113
Cosmetics applications of microalgae,
       82–83
COVID-19 pandemic, 8
Crude oil, 46
Cutting-edge technologies, 124
Cyanobacteria, marine, 51–52
    potential applications of, *52*
Cylindrospermum sp., 52
Cynobium, 51
Cystoseira spp., 107
Cytochrome P450, 62

D

Dark biotechnology, 2
Deuteromycota, 61
DHA, *see* Docosahexaenoic acid
Diesel biodegradation, marine bacteria in,
       46–47
Disease resistance and plant immunity, 108, *109*
Docosahexaenoic acid (DHA), 81
Dunaliela saline, 81
Dunaliella terticola, 81

E

EBM, *see* Ecosystem-based management
Economic viability, 130
Ecosystem architects and emerging pathogens,
       marine nematodes as, 94–95
Ecosystem-based management (EBM), 129
Edible seaweeds, 23, 109
    carotenoids, 31–32
    seaweed phlorotannin, 30–31
    seaweed polysaccharides, 24
        alginic acid, 26–27
        carrageenan, 27
        fucoidans, 24
        laminarin, 24–26, *25*
        ulvan, 27–30
    sterols, 32
Effective waste management strategies, 12
Eicosapentaenoic acid (EPA), 81
Endoparasitic nematodes, 95
Enhanced reputation, 130
Environmental monitoring
    and bioindication, 95–96
    and remediation, 128
Enzymes
    isolated from marine fungi, 61–62
    marine-based, 97
EPA, *see* Eicosapentaenoic acid
Ethanol, 33
Ethyl acetate, 33
Europe, blue bioeconomy sector in, 2–3
Exopolysaccharides, 113

F

Facultative marine fungi, 60
Filamentous fungi, 65
Filtration, 78
Fisheries management, marine nematodes in,
       93–94
Fish farming practices, intensive, 41
Fish gelatin
    biomedical applications of, 115
    food industry applications of, 114–115
    general extraction protocol of, 114
Fishing industry, 3
Floats, 20
Flotation, 78–79
Food industry, applications of microalgae in, 80–81
Free-living marine nematodes, 92, 93
Frond, 20
Fucoidans, 24
L-Fucose, 24
Fucoxanthin, 32, *32*
Fucus spiralis, 107
Functional food industry, roles of bioactive
       marine compounds in, 109–110
Fungi, general facts about, 60
Fungi, marine, 60
    application of, 61–67
    classifications, origin, and habitats of, 60–61
    enzymes isolated from, 61–62
    future directions, 67–68
    habitats of, 60
    heavy metals bioabsorption properties of, 65
    polycyclic aromatic hydrocarbons (PAHs)
          degradation properties of, 62–64, *63*
    surface-active proteins (hydrophobins) of,
          65–67

G

Gas-filled bladders/floats, 20
Gelatin, 113
    biomedical applications of, 115
    food industry applications of, 114–115
    general extraction protocol of, 114
Gelidium crinale, 107
Gene prospecting, 10
Glucuronic acid, 65
Gordonia, 49
Gracilaria ornata, 108
Gray biotechnology, 1, **2**
Green biotechnology, 1, **2**
Green seaweeds, 22

H

HDPE-degrading bacteria, *see* High-density
          polyethylene-degrading bacteria

Heavy metal bioremediation, role of marine
        bacteria in, 43–46
Heavy metals bioabsorption properties of marine
        fungi, 65
High-density polyethylene (HDPE)-degrading
        bacteria, 49
High-molecular-weight (HMW) PAHs, 62
Hippoglossus stenolepis, 95
HMW PAHs, *see* High-molecular-weight PAHs
Holdfast, 20
Hydrophobins, 65–66
Hysterothylacium sp., 98
    H. reliquens, 98

I

Industrially useful enzymes, 40–41
Industries, applications of upcycling across,
        125–126
Inflammation, 112
Intensive fish farming practices, 41
International Waterski and Wakeboard
        Federation, 5
Iota-carrageenan, 27

K

Kappa-carrageenan, 27
Kappaphycus alvarezii, 106

L

Lactic acid bacteria, 44
Lactobacillus plantarum, 44–45
Lambda-carrageenan, 27
Lamina or Blades, 20
Laminaria sp., 106
    L. japonica, 32
Laminarin, 24–26, *25*
LC-PUFAs, *see* Long-chain polyunsaturated
        fatty acids
Lipases, 41
LMW PAHs, *see* Low-molecular-weight PAHs
Long-chain polyunsaturated fatty acids
        (LC-PUFAs), 81
Low-molecular-weight (LMW) PAHs, 62
Lyngbya, 51

M

Major sectors of blue bioeconomy, 1, *13*
    biomedical industry, 3–4
    biotechnology, 1–2
    fishing industry, 3
    future directions, 12
    marine and coastal tourism, 4
        coastal tourism, 5

        maritime parks, tours to, 5
        scuba diving, 4
        spearfishing, 4–5
        water skiing, 5
        wildlife mammal watching, 5
    marine bioprospecting, 8
        bionic prospecting, 10
        chemical prospecting, 9–10
        gene prospecting, 10
    marine mining and mineral resources,
        10–11
    marine renewable energy, 5
        biomass energy, 7–8
        marine current energy, 7
        ocean thermal energy, 8
        ocean wave/current energy, 7
        offshore solar energy, 6
        offshore wind, 6
        salinity gradient energy, 8
        tidal range energy, 6–7
    marine waste management, 12
    world scale, blue bioeconomy on, 2–3
Marine and coastal tourism, *see* Coastal tourism
Marine bacteria, *see* Bacteria, marine
Marine bioresources, *see* Bioresources, marine
Marine biotechnology applications in sustainable
        aquaculture, 127
Marine carbonate sediments, 11
Marine current energy, 7
Marine cyanobacteria, *see* Cyanobacteria, marine
Marine debris, defined, 122
Marine ecosystems, 9
Marine enzymes
    advantages and disadvantages of using, *42*
    major properties and advantages of, *40*
Marine fish by-products, gelatin separated
        from, 113
    biomedical applications of fish gelatin, 115
    food industry applications of gelatin,
        114–115
    general extraction protocol of fish gelatin, 114
Marine fungi–based hydrophobin, applications
        of, *68*
Marine fungi, *see* Fungi, marine
Marine litter management approaches,
        124, *124*
Marine macroalgae, *see* Seaweeds
Marine microalgae, *see* Microalgae, marine
Marine microorganisms, *see* Microorganisms,
        marine
Marine mining and mineral resources,
        10–11
Marine nematodes, *see* Nematodes, marine
Marine organisms, pharmaceutical discoveries
        from, 127
Marine resource management, circular supply
        chains in

benefits of
    climate resilience, 130
    economic viability, 130
    enhanced reputation, 130
    resource conservation, 130
key components of
    community engagement, 129–130
    resource recovery and recycling, 129
    sustainable fishing practices, 128–129
    traceability and transparency, 129
Marine waste management, 12, 122
    marine plastic management strategies,
        124–125
    plastic as marine contaminant, 122–123, *123*
Maritime parks, tours to, 5
Marylynnia sp., 99
Medicinal value of marine bioresources, 110
    antibiotics, 111
    anticancer compounds, 111–112
    anti-inflammatory agents, 112
    antiviral properties of marine organisms, 111
    cardiovascular medications, 112–113
Methanol, 33
Microalgae, cultivation of, 74
    closed systems, 75–76
    open systems, 74–75
Microalgae, marine, 73
    applications of, 80
        agricultural applications, 81–82
        biofuel production, 84–85
        cosmetics applications, 82–83
        in food industry, 80–81
        pharmaceutical applications, 82
        wastewater treatment applications, 83–84
    future directions, 85
Microalgae harvesting techniques, 76, *76*, 77
    advantages and disadvantages of, **77**
    biological methods
        bioflocculation, 79
        microbial lysis, 80
        predation, 79
    chemical methods, 79
        auto-flocculation, 79
        coagulation, 79
    physical methods, 78
        centrifugation technique, 78
        filtration, 78
        flotation, 78–79
        sedimentation, 78
Microbial lysis, 80
Microorganisms, marine, 39
    future prospectives, 52–53
    major types of, 39–50
'Mismanaged' plastic waste, 12
Monodictys pelagica, 65
Mycosporine, 113
Mycosporine-like amino acids, 113

N

Nannochloris sp., 84
Nannochloropsis spp., 73
National Oceanic and Atmospheric
        Administration (NOAA), 3
Nematodes, marine, 92
    applications of, 93, *93*
        in aquaculture and fisheries management,
            93–94
        bioremediation potential in polluted
            environments, 97–99
        biotechnological potential, 97
        as ecosystem architects and emerging
            pathogens, 94–95
        in environmental monitoring and
            bioindication, 95–96
        nutrient cycling in marine ecosystems,
            contribution to, 96–97
        in seafood industry and aquaculture, 95
    biodiversity and taxonomy of, 92–93
    future directions, 100
Neochromadora peocilosoma, 99
New plant fertilizer products, marine plant
        resources in, 105, *106*
    aquaponics and hydroponics applications,
        107–108
    biodegradable mulching materials, 108
    biostimulation properties reported from
        marine plant fertilizers, 106
    disease resistance and plant immunity, 108, *109*
    soil salinity stress, alleviation of, 107
    stimulating root growth and nutrient uptake, 108
NOAA, *see* National Oceanic and Atmospheric
        Administration
Novosphingobium, 49
Nutrient cycling in marine ecosystems, contribution
        of marine nematodes to, 96–97
Nutrient cycling within aquaculture ponds, 94

O

Ocean thermal energy, 8
Ocean wave/current energy, 7
Offshore solar energy, 6
Offshore wind, 6
Omega-3 fatty acid-rich fish oil, 105
Oncholaimus campylocercoides, 99
Oncorhynchus spp., 95
Osmotic energy, 8
Oxygenation, 97

P

PADI, *see* Professional Association of Diving
        Instructors
Padina gymnospora, 107

PAHs, *see* Polycyclic aromatic hydrocarbons
Paradendryphiella salina, 67
PBRs, *see* Photobioreactors
PCL, *see* Polycaprolactone
PE, *see* Polyethylene
Penicillium sp.
   P. ilerdanum, 63
   P. roseopurpureum, 67
Petroleum and diesel biodegradation, marine
   bacteria in, 46–47
Pharmaceutical applications of microalgae, 82
Pharmaceutical discoveries from marine
   organisms, 127
Phlorotannin, *see* Seaweed phlorotannin
Phormidium sp., 52
Phosphate, 65
Photobioreactors (PBRs), 74, 75
Physical microalgae harvesting methods
   centrifugation technique, 78
   filtration, 78
   flotation, 78–79
   sedimentation, 78
Planktonic cyanobacteria, 51
Plastic as marine contaminant, 122–123, *123*
Plastic debris, 12
Plastic degradation, role of marine bacteria in,
   47–50, *48*
Plastic-degrading agents, applications of marine
   bacteria as, **50**
Plastic management strategies, 124–125
Plastisphere, 49
Polluted environments, bioremediation potential
   of marine nematodes in,
   97–99
Polyamide, 47
Polycaprolactone (PCL), 49
Polycyclic aromatic hydrocarbons (PAHs),
   62–64, *63*, 99
Polyethylene (PE), 47, 64
Polyethylene terephthalate, 47
Polyphenols, 113
Polypropylene, 47
Polysaccharides, 65; *see also* Seaweed
   polysaccharides
Polystyrene (PS), 47, 49
Polyurethane, 47
Polyvinylchloride, 47
Predation, 79
Probiotics, 42
Prochlorococcus, 51
Professional Association of Diving Instructors
   (PADI), 4
Propagules, 21
Proteases, 41
PS, *see* Polystyrene
Pseudanabaena sp., 52

Pseudomonas sp., 49
   P. aeruginosa, 46
   P. chengduensis, 46
Pseudoterranova spp., 95
Purple biotechnology, 2, **2**

R

Raceway ponds, 74, *74*
Reactive oxygen species (ROS), 110
Recycling, 124
Red biotechnology, 1, **2**
Red seaweeds, 22, *28*
Renewable energy, marine, 5
   biomass energy, 7–8
   marine current energy, 7
   ocean thermal energy, 8
   ocean wave/current energy, 7
   offshore solar energy, 6
   offshore wind, 6
   salinity gradient energy, 8
   tidal range energy, 6–7
Reproduction in seaweeds, 21
   asexual reproduction, 21
      propagules, 21
      spores, 21–22
      thalli fragmentation, 21
   sexual reproduction, 22
Resource conservation, 130
Resource recovery and recycling, 129
Rhabditis (*Pellioditis*) *marina*, 99
ROS, *see* Reactive oxygen species

S

Salinity, 107
Salinity gradient energy, 8
Samuelson, Ralph, 5
Sargassum sp.
   S. fusiforme, 32
   S. vulgare, 106, 108
Scenedesmus sp., 84
Schizochytrium sp., 52, 84
Schizophyllium commune, 64
Scopus database, 3
Scuba diving, 4
Seafloor massive sulfide, 11
Seafood industry and aquaculture, effects of
   nematodes in, 95
Sea vegetables, *see* Edible seaweeds
Seawater, 7
Seaweed phlorotannin, 30–31, 33
Seaweed polysaccharides, 24
   alginic acid, 26–27
   carrageenan, 27
   fucoidans, 24

laminarin, 24–26, *25*
ulvan, 27
   agriculture, 29
   biomedical applications, 29–30
   cosmetics, 30
   environmental applications, 30
   food industry, 29
Seaweeds, 19
   asexual reproduction in, 21–22
      propagules, 21
      spores, 21–22
      thalli fragmentation, 21
   bioactive compounds reported from edible
      seaweeds, 23
      carotenoids, 31–32
      seaweed phlorotannin, 30–31
      seaweed polysaccharides, 24–30
      sterols, 32
   botanical facts about, 19–20
   classification of, 22
   composition of, 22–23
   future directions, 34
   general facts about, 19
   life cycle of, *23*
   limitations related to bioactive compounds
      isolated from, 33
      bioactive properties, validation of, 33
      heavy metals, presence of, 33
      improving the yield and characteristics of
         seaweed-based product, 34
      purified compounds, structural
         characterization of, 33
      solvents and high cost, toxic nature of, 33
   sexual reproduction in, 22
   structural motifs of, *25*
   structure of, *20*
   vegetative reproduction in, 21
Sedimentation, 78
Sexual reproduction in seaweeds, 22
Shellfish, 3
Soil salinity stress, alleviation of, 107
Solar thermal systems, 6
Spearfishing, 4–5
Spectrophotometer-based methods, 33
Spermatia, 22
Spirinia sp.
   S. gerlachi, 99
   S. parasitifera, 99
Spirulina sp., 52
   S. platensis, 52, 81
Spores, 21–22
Sterols, 32, 33
Stipe, 20
Sulfated polysaccharides, 105
Surface-active agents, 46
Surface-active proteins of marine fungi, 65–67

Surfactants, 46
Sustainability and marine biotechnology, 126
   environmental monitoring and remediation,
      128
   marine biotechnology applications in
      sustainable aquaculture, 127
   pharmaceutical discoveries from marine
      organisms, 127
Sustainable aquaculture, marine biotechnology
      applications in, 127
Sustainable blue economy strategy, 11
Sustainable fishing practices, 128–129
Synechococcus, 51
Synechocystis, 51
Synthetic agricultural agents, removal of, 9

T

Talaromyces pinophilus, 67
Terschellingia longicaudata, 99
Thalli fragmentation, 21
Thallus, 20
Thick-wall zygospores, *see* Zygotes
Tidal range energy, 6–7
Tidal stream energy, 7
TK, *see* Traditional knowledge
Traceability and transparency, 129
Traditional knowledge (TK), 129–130
Trichoderma harzianum, 67
Triticum durum L., 107

U

Ulva lactuca, 28, 107
Ulvan, 27
   in agriculture, 29
   in biomedical applications, 29–30
   in cosmetics industry, 30
   in environmental applications, 30
   in food industry, 29
Ulvanobiose U3S, 28
Ulvanobiuronic acid, 28
Undaria pinnatifida, 32, 107
Upcycling of marine resources and by-products
   applications of, 125–126
   challenges associated with, 126
   community engagement and education, 126
   environmental stewardship and waste
      reduction with, 126
   principles of, 125

V

Vegetative reproduction in seaweeds, 21
Violet/purple biotechnology, 2, **2**

W

Waste management, marine, 12
Wastewater treatment applications of microalgae,
    83–84
Water skiing, 5
White biotechnology, 1, **2**
Wildlife mammal watching, 5
World scale, blue bioeconomy on, 2–3

X

Xanthomonas campestris, 108
X-ray diffraction analysis, 64
Xylanases, 41

Y

Yellow biotechnology, 1, **2**

Z

Zea mays, 106
Zoospores, 21, 22
Zygomycota, 61
Zygospores, 22
Zygotes, 22